High Strength Support for Soft Surrounding Rock in Deep Underground Engineering

Qi Wang · Bei Jiang · Shucai Li

High Strength Support for Soft Surrounding Rock in Deep Underground Engineering

Theory and Key Technology

Springer

Qi Wang
School of Qilu Transportation
Shandong University
Jinan, China

Bei Jiang
China University of Mining
and Technology-Beijing
Beijing, China

Shucai Li
School of Qilu Transportation
Shandong University
Jinan, China

ISBN 978-981-15-3846-9 ISBN 978-981-15-3844-5 (eBook)
https://doi.org/10.1007/978-981-15-3844-5

This Springer imprint is published by the registered company Springer Nature Singapore Pte Ltd.
The registered company address is: 152 Beach Road, #21-01/04 Gateway East, Singapore 189721, Singapore

Preface

The coal industry is experiencing a rapid development with the continuous increasing of global energy demand. As the shallow coal resources tend to dry up, coal mining develops to the deep. We are often faced with challenging problems in deep excavations of underground engineering with soft surrounding rocks, such as large deformation, floor heave and failure of supporting components. The existing support methods do not meet the surrounding rock control requirements. So, it is very crucial to conduct research and development on new theories and technologies for surrounding rock control.

In this book, a control concept of "high strength, pressure yielding and integrity" is put forward for surrounding rock control in the deep underground engineering with soft rocks; and a new high strength support system is developed and its key technologies, design principles and construction methods are introduced.

The layout of monograph is as follows:

Chapter 1 gives general introduction followed by an overview of surrounding rock support in deep roadways. This chapter mainly introduced the theory and technical development of surrounding rock support in deep roadways, current situation of U-steel arch support and development status of the confined concrete support system.

Chapter 2 gives a detailed description of confined concrete support system. Through the on-site investigation, the failure mechanism of the bearing structure of deep soft rock roadways is clarified; a coupling support concept is proposed; and a confined concrete support system is established which consists of an internal high-strength bearing layer, an intermediate filling adjustment layer and an external anchoring self-supporting layer.

Chapter 3 detailed introduces the mechanical properties of basic confined concrete components. Through the numerical test and experiment methods, systematic studies are made on the axial bearing capacity of confined concrete short columns and the reinforcing mechanism of components with grouting holes. A pure bending numerical analysis model is established for the casing joint. Through the model, the deformation and failure mechanism and the bearing mechanism of the casing joint are clarified; the parameters are analyzed; and the influence law of different parameters (concrete strength f_{cu}, k, casing length l, casing wall thickness t, clearance δ, etc.) are clarified on the mechanical parameters such as ultimate bending moment, critical curvature increment and equivalent flexural rigidity. All experimental data are the foundation for the further discussion.

Chapter 4 mainly introduces the calculation theory of confined concrete arches. The calculation and analysis model is established for the confined concrete arch casing joint, and an analysis is made on the mode of action and mechanical behavior of the casing joint. Based on the calculation analysis model, the internal force calculation formulas are derived from straight-wall semi-circular arches and circular arches with unequal stiffness and arbitrary section numbers. The influence of different parameters is clarified on the internal force of the arch, such as load q_1, lateral pressure coefficient λ, arch flexural rigidity EI, joint equivalent stiffness ratio μ, joint positioning angle α, and height-diameter ratio κ. Combined with the calculation method of the ultimate bearing capacity of confined concrete components, the ultimate bearing capacity of arches is compared and analyzed with different shapes of the cross-sections. All of the theories are the foundation to analyze the mechanism and the phenomenon of confined concrete.

Chapter 5 mainly introduces the bearing mechanism of confined concrete arches through experimental study. To further understand and master the bearing mechanism of confined concrete arches, a large mechanical test system is designed and developed for the full size confined concrete arches of underground engineering. The comparison tests are carried out on the full-scale arches of SQCC, CCC and U-steel; the study is made on the influence factors and laws of the mechanical properties of arches; and the influence mechanism of concrete core strength, lateral pressure coefficient and steel pipe wall thickness is defined on the ultimate bearing capacity of the arches.

Chapter 6 mainly introduces the field application research of the confined concrete support system. Based on the engineering backgrounds of the typical soft rock mine Liangjia Coal Mine, the field application of the confined concrete support system is carried out with the on-site monitoring and the numerical experimental research, the failure mechanism of the bearing structure of the soft rock roadway and the control mechanism of the surrounding rocks are clarified.

Chapter 7 mainly introduces the field application research of the confined concrete support system. Based on the engineering backgrounds of the typical high stress kilometer deep mine roadway in Zhaolou Coal Mine, the field application of the confined concrete support system is carried out with the on-site monitoring and the numerical experimental research, the failure mechanism of the bearing structure of the kilometer deep mine roadway and the control mechanism of the surrounding rocks are clarified.

In this book, chapters are carefully developed to cover the description of confined concrete support system, mechanical properties of basic confined concrete components, calculation theory of confined concrete arches, bearing mechanism of confined concrete arches through experimental study and field application research of the confined concrete support system. This book is written for researchers of rock mechanics engineering, mining engineering, and safety engineering.

Finally, we would like to thank the following students for their outstanding work, including Hongke Gao, Bohong Liu, Shuo Xu, Zhongxin Xin, Yue Liu, Yuchi Xiao, Yue Wang, Zhenhua Jiang, Peng Zhang, Weiteng Li, Dechao Wang, Rui Pan, Xing Shao, Hengchang Yu, Qian Qin and Yingcheng Luan.

Jinan, China

Qi Wang
Bei Jiang
Shucai Li

Acknowledgements

This book was financially supported by the National Natural Science Foundation of China (grant number 51674154, 51874188, 51704125), the Natural Science Foundation of Shandong Province, China (grant number 2019SDZY04, 2017GGX30101, 2018GGX109001, ZR2017QEE013), the Program for Youth Innovative Research Team in University of Shandong Province, China (grant numbers 2019KJG013), The Research Fund of The State Key Laboratory for GeoMechanics and Deep Underground Engineering, CUMT (grant number SKLGDUEK1717) and Young Scholars Program of Shandong University (grant number 2018WLJH76).

Contents

About the Authors

Dr. Qi Wang is currently a professor at Shandong University in China. He obtained the Ph.D in geotechnical engineering from Shandong University, and he is now the vice-chairman of Soft Rock Engineering Sub-society of Chinese Society for Rock Mechanics & Engineering (CSRME) and the director of the Shandong Province Engineering Laboratory of Safety Control for Underground Space. He has won the outstanding mid-aged experts in Shandong Province, and was awarded the Youth Science and Technology Award by CSRME and China Coal Society (CCS). He took charge of 3 National Natural Science Foundation of China (NSFC), 8 provincial research projects and 19 projects from enterprises and institutions. His research focuses on the high strength support theory and technology for underground engineering. Up to now, he has published 66 SCI or EI papers as the first or corresponding author and hold 35 first Chinese invention patents. He has been awarded 13 prizes at the provincial and ministerial level, including the First Prize of the Technology Invention in Shandong Province, the First Prize of the Science and Technology Progress in Shandong Province, and China Invention Patent Excellence Award, *etc*.

Dr. Bei Jiang obtained the Ph.D. in geotechnical engineering from Shandong University in 2016. She works at China University of Mining & Technology, Beijing, and she is now the standing director of Soft Rock Engineering Sub-society of Chinese Society for Rock Mechanics & Engineering (CSRME). She took charge of more than 7 projects such as National Natural Science Foundation of China (NSFC), Shandong Provincial Natural Science Foundation, China Postdoctoral Science special Foundation. Her research focuses on the surrounding rock failure mechanism and the control theory of underground engineering with complex conditions. She has published more than 60 papers indexed by SCI or EI in related fields. And she has 52 Chinese invention patents. She has been awarded 6 prizes at the provincial and ministerial level, including the First Prize of Science and Technology by China Coal Industry, Chinese Association of Construction Enterprises, and CSRME, *etc*.

Dr. Shucai Li is an Academician of the Chinese Academy of Engineering. He works as professor in Shandong University, and he is now the vice-principal of Shandong University and the vice-chairman of Chinese Society for Rock Mechanics & Engineering (CSRME). He is the chief scientist of National 973 Program of China, the distinguished professor of Chang Jiang Scholars and the winner of the National Science Fund for Distinguished Young Scholars. He won the Highest Science and Technology Award of Shandong Province, and took charge of more than 20 national, provincial and ministerial projects including key projects of National Natural Science Foundation of China (NSFC), *etc*. His research focuses on the stability analysis and safety control of underground engineering. Up to now, he has published more than 200 papers included in SCI or EI, and has more than 40 invention patents as the first authorized. He has won 3 second prizes of National Science and Technology Progress of China.

Chapter 1
Introduction

1.1 Overview

The energy demand of the world continues to increase with the development and recovery of the global economy, and therefore the coal industry is experiencing a rapid development to meet the growing demand for coal. As the most abundant fossil fuel in the world, coal resource is rich. According to the statistics from the BP Statistical Review of World Energy (2019), the total proved coal reserves are 1.055 trillion tons worldwide in 2018, and about three times the oil and natural gas reserves. The world coal production was 3.916 billion tons in 2018, increased 4.3% over the previous year; the global coal consumption in 2018 was 3.772 billion tons of oil equivalent, increased 1.4% over the previous year. Global energy demand will grow by about one-third by 2040, and the demand for coal will reach 270 million tons of oil equivalent. So coal will remain the main source of energy that supports the world's economic and social development in the future. Along with the increasing demand for coal energy and the increasing mining intensity in the world, the coal mining depth has reached to over 1000 m in many countries such as Switzerland, Canada, Australia, China and South Africa. With the depletion of shallow coal resource, it is imperative for coal mining to move further into the deep (He et al. 2005; He 2014; Kang et al. 2019; Xie et al. 2019).

The mining in deep coal mines is affected by complex stress fields in deep strata, and will face the engineering rock mechanics problem which differs completely from that of the shallow mining. The engineering rock mass changes from a linear rock mechanics behavior in the shallow to a nonlinear rock mechanics behavior in the deep; and there are many roadways with supporting difficulties such as deep high-stress roadways, soft rock roadways, roadways affected by strong dynamic pressure, and extremely broken roadways with loose fractures of surrounding rocks. The deformation speed and amount of roadways increase by 20–30% for every

Q. Wang et al., *High Strength Support for Soft Surrounding Rock in Deep Underground Engineering*, https://doi.org/10.1007/978-981-15-3844-5_1

additional 100 m, starting from 500 m under the same conditions (Zhang et al. 2013). According to incomplete statistics (Wang and Gao 2013; Sun et al. 2017; Zhang et al. 2019b), the actual repair rate of deep roadways is over 90%, and the roadway repair rate at the depth of 1000 m is about 3–15 times that at the depth of 500–600 m. Deep mining has many challenges such as higher maintenance cost, poor operation of the production system, insufficient transportation capacity and potential safety hazards; and it always causes safety accidents such as the deformation of surrounding rocks, rib spalling and roof falling. The safety accidents account for over 40% of the total number of mine accidents and the death toll accounts for over 50% of the death rate per million tons in mines.

At present, relevant researchers have begun to pay attention to the support problem of deep roadways with soft rocks, and a large amount of material and manpower are put forward into the related research every year. However, due to the influence of complex geological conditions on the deep high-stress soft rock roadway, many problems occur frequently such as large deformation, floor heave and the failure of the supporting components of the roadway. The existing supporting system is hard to meet the surrounding rock control requirements. Therefore, the research on the theory and technology of deep surrounding rock control is still a very important and also very difficult research subject.

In the development of the mine roadway support, it has taken over 60 years that the support concept has evolved to the current flexible support (New Austrian Tunnelling Method) (Rabcew 1960) from the early rigid support, along with the increase of mining depth and the continuous improvement of roadway support technologies. In the later decades, the conventional flexible support (anchor spray) can still effectively control the stability of the roadway since most of the roadways haven't reached the burial depth of 1000 m. As the result, the rigid support is considered as a backward support form and should be abandoned. At the same time, New Austrian Tunnelling Method is often simply understood as a flexible support. However, more and more research results show that the rigid support with a certain high strength is also essential for the stability of the surrounding rock of the chambers just like the flexible support corresponding to the anchor. The two seemingly conflicting views are actually not contradictory at all. In fact, New Austrian Tunnelling Method is not a simple flexible support. Its essence is to fully mobilize the bearing capacity of the surrounding rock through the support function, and to have the surrounding rock as the main structure to bear the upper pressure. The purpose of high-strength support is not simply to provide rigid support. The essence is to provide greater radial force to the surrounding rock around the roadway, so that the surrounding rock can form a strong and stable load-bearing structure. Therefore, the study of new support systems is of great significance for the surrounding rock control of deep roadways.

The research on the new control theory and technology of deep surrounding rocks is the key to ensure the safety and the high efficiency of deep mining. It is of great significance to the development of infrastructure construction and the improvement of subject research content. This book will carry out a systematic and in-depth research on this issue from many aspects.

1.2 Development and Research Status of Surrounding Rock Support in Deep Roadways

1.2.1 Overview of the Theory Development of Surrounding Rock Support in Deep Roadways

The roadway excavation destroys the original equilibrium state of the surrounding rock. Before the surrounding rock reaches a new equilibrium state, it is inevitable that the surrounding rock undergoes a certain amount of deformation. The large deformation of the surrounding rock often leads to various underground engineering accidents and causes considerable difficulties and challenges in the design and construction of underground engineering. The excessive deformation control of surrounding rocks under the ground pressure has become an important research direction in geotechnical engineering. The classical pressure theory represented by Heim, Langkin and Kinnick Theory developed in the early 20th century (Yu et al. 1983; Zheng 1988) believes that the pressure acting on the supporting structure is the quality of the overlying strata. With the increase of excavation depth, it is found that classical pressure theory is inconsistent with the actual in many aspects, so the theory of slump arches is developed, which is represented by the theories of Taisha and Platts (Zhou et al. 2002; Jizhao 1981). The Austrian engineer L.V. Rabcewic summarized and proposed New Austrian Tunnelling Method (NATM) in the 1960s based on the tunnel construction experiences for many years in Austria. Consisting of flexible and active support methods, NATM has certain guiding significance for soft rock roadways and is widely used in the design and construction of underground engineering (Brown 1990; Barton and Grimstad 1994; Liu 2000). Yamato and Sakurai Sakusuke of Japan proposed the strain control theory of surrounding rock support, and Salamon et al. proposed the theory of energy support (Zheng 1988). Dong (1997, 2001) proposed the theory of surrounding rock loose circle. The viewpoint of this theory is summarized as follows: the larger the loosing ring of the roadway is, the larger the deformation of the surrounding rock is inward the hollow roadway, and the more difficult the support is. The purpose of the support is to reduce the loosing ring of the surrounding rocks and prevent the ring from developing into harmful deformation. He et al. (2000) proposed the soft rock engineering mechanics support theory by combining the engineering geology with the modern large deformation engineering mechanics support theory.

At present, numerical calculation methods are becoming more and more mature, such as finite element method, boundary element method and discrete element method; and a great number of the calculation softwares based on this theory have emerged and are widely known, such as ADINA, NOLM, FINAL, UDEC, SAP and FLAC. The combination of the software and some support theories has been widely used in underground engineering support (Aleksander et al. 2019; di Prisco et al. 2018; Zhang et al. 2019a).

Since the support concept of NATM has become the guiding principle in the design of roadway support, flexible supports such as anchors have played an important role in the support of shallow and medium-deep roadways. However, in the support of the deep soft rocks, the engineering practice of a large number of mining areas with outstanding problems shows that: At present, the bolt support technology is still highly dependent on the occurrence conditions of the surrounding rock. In the broken soft rock roadway where the joint fissure is extremely developed, the bolt support is difficult to form a stable and effective bearing structure in the shallow surrounding rock of the roadway. The support is hard to fully exert its bearing capacity and cannot effectively control the deformation of such roadways. With the increasingly complex geological conditions of the roadway, it is not enough to ensure the overall stability of the roadway by adopting a single flexible support technology for the surrounding rock of the deep soft rock roadway.

Therefore, since the 1990s, the combined support technologies of anchor belt nets and cables, and anchor net trusses (Witthaus et al. 2006; Bawden and Tod 2003; Liu et al. 2005; Brummer and Swan 2001; Holmgren 2001; Kaiser and Tannant 2001; Spearing 2001; Villaescusa 1999; Fuller 1999) have been developed gradually. Secondary support measures with high-strength and high-rigidity are adopted to support the surrounding rock stress jointly on the basis of the initial pressure-yielding support of surrounding rocks by bolts. The purpose of the combined support is to increase the support strength to maintain a long-term stability of the roadway. The support measures are mainly composed of bolts, anchor cables, grouting and arches or concrete lining structures to form different combined supports. The main support technologies include anchor net spray + key part anchor cable support, anchor net spray + U-steel arch support, U-steel collapsible arch injection support with reserved deformation, and anchor net spray + confined concrete arch support.

The early roadway support was mainly based on the classical pressure theory and the collapsed arch theory, and the shed-type rigid support was mostly adopted. Then with the introduction of NATM, the flexible support represented by the anchor net spray support was widely applied and became the main way of roadway support in shallow coal mines (Xie et al. 2010); however, as the depth of roadways gradual increases, the problem of insufficient supporting strength is exposed on the anchor net spray support. The support is hard to achieve the desired support effect under the complex conditions such as high ground stress, soft rocks and fault fracture zones, and even safety accidents such as roof falling may occur (Liu et al. 2011; Lu and Wang 1991). Therefore, the combined support method represented by the U-steel arch + anchor net spray has become the mainstream, and once achieved good results.

1.2.2 *Current Situation of Traditional Support*

At present, the traditional support represented by U-steel arches is commonly used in deep soft rock roadways. Acting directly on the surrounding rock surface of the roadway, the passive radial support force provided by the U-steel arch balances the deformation pressure of the surrounding rock and restrains the deformation of the surrounding rock. U-steel was first developed by the German company Heitzmann in 1932. With excellent cross-sectional geometric parameters, it is easy to realize lap joint shrinkage and therefore, widely used in coal mine roadways. The application rate of U-steel arches reached 90% in major coal-producing areas in Germany from 1972 to 1977 and the serialization production of the arches was made (Hou 1989). Today it is still the most commonly used support form for deep soft rock roadways.

1.2.2.1 Theoretical and Experimental Research

In the research of the U-steel arch supporting theory, Mitri and Hassani (1990) made the numerical simulation on the three kinds of arches widely used in coal mine roadways with numerical simulation technology in 1989, based on nonlinear finite element analysis. In the study, the strength and stiffness characteristics and the load distribution characteristics of the arch were discussed; and the effects of diameter, shape, arch angle and cross-sectional dimensions of the arch were analyzed. (You 2000a, b) studied the calculation theory of the metal arch of the roadway, and systematically expounded the calculation principle, the internal force calculation, force transmission characteristics, shrinkage analysis and stability analysis of the U-steel shrinkable arch.

As for the tests on mechanical properties of U-steel arches, Jukes et al. (1983) studied the influence of the arch spacing on the overall stability of the supporting system through a large number of experiments; and he obtained the relationship between the optimal arch spacing value and the roadway in the study. Khan et al. (1996) of Canada conducted experiments on five different types of roadway arches to study the force characteristics and failure modes of the roadway arches in 1995; and he made a conclusion that the strength and rigidity of the steel support in the roadway depend largely on the external force constraint on the arch legs. Tamada et al. (2006) verified the correctness of the shear strength calculation formula by testing the ultimate shear strength of U-steel. Liu et al. (2011) summarized seven failure types of U-steel arches based on deformation characteristics and mechanical reasons, and proposed the concept and connotation of buckling of U-steel arches.

1.2.2.2 Problems of the Traditional Support in Field Application

At present, the U-steel arch is commonly used for deep high-stress soft rock roadways and chambers. However, as a traditional support form, it has the following problems in the surrounding rock support:

(1) Insufficient supporting strength. The U36 steel arch is always yielded and broken in many deep mining areas. With the steel content of 35.87 kg/m, the U36 steel for mining has a large steel consumption and a low-cost performance.

(2) Low working resistance. Since the U-steel is connected by a lap joint, the supporting reaction force of the U-steel arch is related to the frictional resistance of the lap joint, rather than the compressive strength of the U-steel itself. In China, U-steel arches generally are insufficient in locking force and relatively low in working resistance at the lap joint. Numerous tests and measured results show that the supporting force of the U-steel arch is not fully utilized, and its working resistance reaches only about 50% of its ultimate bearing capacity.
(3) Low material utilization. Due to its unique cross-section, U-steel arches are generally in point and line contact with surrounding rocks, so that the supporting structure is subjected to concentrated or eccentric load. The load deteriorates the stress state of the supporting structure. Therefore, the support capacity is not fully utilized, the material utilization rate is reduced, and the support effect is not guaranteed.
(4) Cannot realize quantitative pressure yielding. The clamp is used as a joint connection form, and a certain amount of pressure can be released. But, it is difficult to grasp and control the pressure amount and the timing of the pressure yielding.

1.2.2.3 Improved Form of the U-Steel Arch

Aiming at the problems in the practice of the existing U-steel arches for mines, corresponding improved methods for the U-steel support forms have been proposed.

Jiao et al. (2013) made an improvement on the traditional U-steel arch, and proposed a support form of metal mesh + post-wall chemical grouting + geo-membrane + U-steel arch, and carried out a field test. Luo and Chang (2009) increased the contact area by filling materials mixed with cement, gypsum and fly ash between the U-steel and the surrounding rock; thereby the mechanical performance of the U-steel was improved; and the U-steel, the filling material and the surrounding rock became an integrated structure to improve the bearing capacity and reduce the roadway deformation. Aiming at the problem of the floor heave in the deep roadways, Zhao et al. (2015) made a numerical analysis on the U-steel model; and he proposed an inverted U-steel to form a sealed arch to improve the overall support effect of the surrounding rock. Tan et al. (2017) proposed a composite support system to control the deformation of the soft rock roadway during excavation and also made a numerical analysis and on-site monitoring. The system was composed of steel mesh, anchor bolt, anchor cable, shotcrete, compressible U-steel, foam concrete damping layer and fractured rock mass cushion.

However, when the cross-sectional bending moment of the U-steel reaches a certain value, the opening cross-section will slowly open up, resulting in a rapid decrease of the bending resistance of the arch and thus a buckling and failure. So far, there is still no universally applicable form of support that addresses the above issues.

Therefore, it is very necessary to find a new type of high strength support to solve the problems of traditional support technology in surrounding rock control.

1.2.3 *Development Status of New High Strength Support*

As a new type of high strength support, the confined concrete is external confined materials filled with concrete. Concrete-filled steel pipe is one of the forms of confined concrete. Concrete core is under three-dimensional compression due to the confined effect of external materials, and its strength is greatly improved. Meanwhile, the external materials are supported by the concrete core, which effectively prevents buckling. Therefore, the two realize the "symbiosis of force". At present, the confined concrete structure has been developed rapidly in engineering, due to its many advantages such as quick construction, high bearing capacity, good ductility, excellent economic performance, fire prevention and earthquake resistance.

1.2.3.1 Bearing Capacity Principle of the New High Strength Structure and Its Application on the Ground

As a new high strength structure, the confined concrete structure has been widely used on the ground due to their good mechanical properties and construction characteristics. In 1879, confined concrete components were used for the first time to prevent corrosion and to bear the pressure of steel pipes in the construction of railway piers in Severn, England. After the 1960s, many countries have formulated the design rules or the specifications for concrete-filled steel pipes. At present, there are relatively mature design codes and technical specification, such as: ACI 318–19 (2019), AISC 360–16 (2016), and AISC LRFD (2011) in the United States, and 《Technical Specification for Concrete Filled Steel Tubular Structure GB50936–2014》, 《Code for Design and Construction of Concrete Filled Steel Tubular Structure CECS28–2012》 in China.

1.2.3.2 The Application of New High Strength Support in Underground Engineering

At present, the application of the new high strength support in underground engineering is generally in the initial stage, and only a few scholars have conducted field tests.

In 1995, Zang and Li (2001) conducted an industrial test on confined concrete arches in the Pingdingshan Mining Bureau. The results show that under the same bearing capacity, the confined concrete arch can save about 30% of steel consumption comparing with the U-steel arch, and reduce the cost by 20% at least. It proves that the confined concrete arch is an ideal support form in underground engineering with high bearing capacity, excellent economic performance and good mechanical properties.

Gao et al. (2010) used confined concrete arches with the connection of casing joint and the technology of installing before grouting in Qianjiaying Mine. Gao et al.

(2016) introduced the jacking grouting method into the confined concrete support to ensure the grouting quality of the concrete core, which is applied in Chagannur Mine.

Liu et al. (2015) researched and developed the D-shaped confined concrete support, which was applied in Pingmei Coal Mine and achieved good control effect of surrounding rock.

Wang et al. (2015a, 2018) researched and developed the U-shaped confined concrete and square confined concrete support for the first time, which were applied in the high stress kilo deep mine (Zhaolou Coal Mine) and the extremely soft rock mine (Liangjia Coal Mine) respectively. Wang et al. (2019) carried out the first application of confined concrete support technology for the super large section tunnel (Longding tunnel).

Throughout the above-mentioned on-site practices, the advantages of confined concrete arches in the deep soft rock roadway, such as high strength, high rigidity and high cost performance, have attracted more and more attention due to a large increase of the kilometer deep mines. The field application has demonstrated the advantages of confined concrete support systems in surrounding rock control.

1.2.3.3 Theoretical and Experimental Research on the Underground Support System of Confined Concrete

1. Experimental Research

Zang and Li (2001) conducted an experimental research of the confined concrete support on a test bench of underground structure. Due to the limitation of the test conditions, only a model test was conducted on the arc-shaped component and a reduced-scale arch. The conclusions are as follows: (1) Confined concrete arches have a high bearing capacity and can meet the support requirements of underground structures with high confining pressure. (2) With high flexibility, confined concrete arches can meet the requirements of underground structure support for soft and expansive surrounding rocks.

Scholars of China University of Mining and Technology (Beijing) (Gao et al. 2010; Liu 2013; Qu 2013) carried out experimental studies on the arc-shaped components, including the comparative analysis of flexural and bearing properties of different component types, such as confined concrete, hollow steel pipe, U36 and I22b. Meanwhile, the mechanical properties of the small-scale arch are tested. The test results show that the bearing capacity of the confined concrete arc component is 2.39 times that of U-steel, 7.63 times that of I-steel, and the bearing capacity of confined concrete arch is 3.17 times that of U-steel arch, under the approximately equal steel weight.

The above analysis shows that there are few experimental studies on confined concrete arches, and the focus of those studies are mainly on the axial compression of short column components, the flexural rigidity of the arc components and the reduced-scale arches. On-site monitoring shows that the arches are generally in a significant bias state, so it is very necessary to test the bias of the confined concrete arches. At present, no special research has been carried out in this area. In addition, very few tests on a whole arch have been carried out in 1:1 scale. Most of them are simulation tests with a single cylinder loading. It is not easy to approach the actual stress state of the arch in the field, nor can it apply load with different lateral pressure coefficients.

2. Theoretical Research

At present, the calculation theory of confined concrete arches for underground engineering support is still in the starting stage, and there is little information available for the relevant theoretical calculations. Many related topics have not been involved such as the bearing capacity of the confined concrete arch on compression bending, the bearing capacity calculation of the joint component and the bearing capacity of the arch under different load modes.

In summary, the coal mine projects are moving further into the deep stratum of complex stress field; the mechanical advantages of the new confined concrete support system of underground engineering have been verified by field tests; however, relevant research is still insufficient, and the systematic theoretical research and experimental support are needed for the new support system.

Based on the development statues, Li et al. (2015, 2016) and Wang et al. (2011, 2015a, b, c, 2016, 2017, 2018, 2019) put forward a variety of high strength support systems for the first time since 2011, such as square confined concrete (SQCC) and U-shaped confined concrete (UCC), which have the advantages of high flexural rigidity, close contact with surrounding rocks and convenient construction of longitudinal connection. The laboratory full scale arch tests are carried out systematically. The calculation models for the internal force of the arches with unequal rigidity and arbitrarily section numbers are established. The design method of confined concrete support is formed, and a complete set of key techniques such as mechanized construction equipment are developed. These methods and techniques are successfully applied in typical mines such as deep high stress, extremely soft rock and thick alluvium, and popularized applied in the large-section traffic tunnels.

1.3 The Main Content of This Book

This book focuses on the key issues of "The Theory and Technology of the Soft Surrounding Rock Control in Deep Underground"; and it conducts systematic research by laboratory experiments, theoretical derivation and numerical simulation. The main research contents are divided into the following five aspects.

1. Research on the New Support System of High Strength

Through the field investigation, the failure mechanism of the bearing structure of deep soft rock roadways is clarified; a coupling support concept of "high strength, pressure yield and integrity" is proposed; and a new support system of high strength is established which consists of an internal high-strength bearing layer, an intermediate filling adjustment layer and an external anchoring self-supporting layer.

2. Study on the Mechanical Properties of Basic Components

Through the numerical simulation and laboratory tests, systematic studies are made on the axial bearing capacity of confined concrete short columns and the reinforcing mechanism of confined concrete components with grouting holes. A pure bending numerical analysis model is established for the casing joint component. Through the model, the deformation and failure mechanism and the bearing mechanism of the casing joint are clarified; the parameters are analyzed; and the influence law of different parameters (concrete strength $f_{cu,k}$, casing length l, casing wall thickness t, clearance δ, etc.) are clarified on the mechanical parameters such as ultimate bending moment, critical curvature increment and equivalent bending rigidity.

3. Theoretical Study on the Calculation of New High Strength Arches

The calculation and analysis model is established for the confined concrete arch casing joint, and an analysis is made on the mode of action and mechanical behavior of the casing joint. Based on the computational analysis model, the internal force calculation formulas are derived for straight-wall semi-circular arches and circular arches with unequal rigidity and arbitrary section numbers. The influence of different parameters is clarified on the internal force of the arch, such as load ql, lateral pressure coefficient λ, arch flexural rigidity EI, joint equivalent stiffness ratio μ, joint positioning angle α, and height-diameter ratio κ. Combined with the calculation method of the ultimate bearing capacity of confined concrete components, the ultimate bearing capacity of arches is compared and analyzed with different shapes of the cross-sections.

4. Experimental Study on the Bearing Behavior of New High Strength Arches

To further study the bearing behavior of confined concrete arches, a large mechanical test system is designed and developed for the 1:1 scale confined concrete arches of underground engineering. The comparison tests are carried out on the full-scale arches of SQCC, CCC and UCC; the study is made on the influence factors and laws of the mechanical properties of confined concrete arches; and the influence mechanism of concrete core strength, lateral pressure coefficient, steel pipe wall thickness and restraint effect coefficient is defined on the ultimate bearing capacity of the arches.

5. Field Application Research on the New Support System of High Strength

Based on the engineering backgrounds of the typical extremely soft rock mine and the kilometer high-stress deep mine, the field application of the new support system of high strength is carried out with the on-site monitoring and the numerical experimental research. The failure mechanism of the bearing structure is analyzed, and the control mechanism of the surrounding rock is clarified.

References

ACI 318–19 (2019) Buiilding Code Requirements for Structural Concrete and commentary, American Concrete Institute, Detroit
AISC 360–16 (2016) Specification for structural steel buildings, American Institute of Steel Construction, Chicago
AISC-LRFD (2011) Load and resistance factor design specification for structural steel buildings, American Institute of Steel Construction, Chicago
Aleksander K, Malgorzata K, Piotr K (2019) Numerical modeling of air velocity distribution in a road tunnel with a longitudinal ventilation system. Tunn Undergr Space Technol 91:103003
Barton Nick, Grimstad Eystein (1994) Rock mass conditions dictate choice between NMT and NTITM. Int J Rock Mech Min Sci 10:39–42
Bawden WF, Tod JD (2003) Optimization of cable bolt ground support using SMART instrumentation. ISRM News J 7(3):10–16
BP (2019) BP Statistical Review of World Energy. UK
Brown ET (1990) Putting the NATM into perspective. J Tunnels Tunneling 7:76–79
Brummer Rk, Swan G (2001) Support and structural use of shotcrete in mines. Undergr Min Methods: Eng Fund Int Case Studies:593–600
CECS 28:2012 (2012) Technical specification for concrete-filled steel tubular structures, Chinese Association for Engineering Construction, China
di Prisco C, Flessati L, Frigerio G, Lunardi PA (2018) Numerical exercise for the definition under undrained conditions of the deep tunnel front characteristic curve. Acta Geotech 13(3):635–649
Dong FT (1997) The theory of supporting with loosening circle in surrounding rock of roadway. J Bolt Support 2:5–9
Dong FT (2001) Theory and application technology of surrounding rock loose circle support. Coal Industry Press, Beijing
Fuller PG (1999) Roof strata reinforcement-achievements and challenges. Rock Support Reinf Pract Min:405–415
Gao YF, Wang B, Wang J (2010) Test on structural property and application of concrete-filled steel tube support of deep mine and soft rock roadway. Chinese J Rock Mech Eng 29(S1):2604–2609
Gao YF, He XS, Chen BH (2016) Study on support technology of concrete-filled steel tubular in roadway with huge thick and rich water soft rock. Coal Sci Technol 44(1):84–89
GB 50936–2014 (2014) Technical Code for Concrete Filled Steel Tubular Structures, Ministry of Housing and Urban-Rural Development of the People's Republic of China, China
He MC (2014) Progress and challenges of soft rock engineering in depth. J China Coal Soc 39(8):1407–1409
He MC, Jing HH, Sun XM (2000) Progress in soft rock engineering geomechanics. J Eng Geology 8(1):46–62
He MC, Xie HP, Peng SP, Jiang YD (2005) Study on rock mechanics in deep mining engineering. Chin J Rock Mech Eng 24(16):2803–2813
Holmgren BJ (2001) Shotcrete linings in hard rock. Undergr Min Methods: Eng Fund Int Case Studies:569–577
Hou CJ (1989) Steel supporting arch in roadway. China Coal Industry Publishing House, Beijing
Jiao YY, Song L, Wang XZ (2013) Improvement of the U-shaped steel sets for supporting the roadways in loose thick coal seam. Int J Rock Mech Min Sci 60:19–25
Jizhao (1981) Taishaji and soil mechanics. J Geotech Eng 3(3):114–115
Jukes SG, Hassani FP, Whittaker BN (1983) Characteristics of steel arch support systems for mine roadways. Min Sci Tech 43–58
Kaiser PK and Tannant DD (2001) The role of shotcrete in hard-rock mines. Undergr Min Methods Eng Fund Int Case Studies 579–592
Kang HP, Xu G, Wang BM, Wu YZ, Jiang PF, Pan JF, Ren HW, Zhang YJ, Pang YH (2019) Forty years development and prospects of underground coal mining and strata control technologies in China. J Min Strata Control Eng 1(1):1–33

Khan UH, Mitri HS, Jones D (1996) Full scale testing of steel arch tunnel supports. Int J Rock Mech Min Sci Geomech 219–232

Li SC, Shao X, Jiang B (2015) Study of the mechanical characteristics and influencing factors of concrete arch confined by square steel set in deep road ways. J China Uni Min Technol 44(3):400–408

Li SC, Wang X, Wang Q, Li WT, Wang FQ, Yang WM (2016) Mechanical property research and failure characteristics of U-type confined concrete arch in deep roadway. Eng Mech 33(1):178–187

Liu CW (2000) Relation between bolting and shotcreting support of soft rock roadway in coal mine and NATM. J China Min Industry 1(1):61–64

Liu GL (2013) Research on steel tube confined concrete supports capability and soft rock roadway compression ring strengthening supporting theory. China University of Mining & Technology, Beijing

Liu B, Yue ZQ, Tham LG (2005) Analytical design method for a truss-bolt system for einforcement of fractured coal mine roofs-illustrated with a case study. Int J Rock Mech Min Sci 42(2):195–218

Liu JZ, Zhang N, Zheng XG (2011) Research on buckling failure mechanism of U type steel support loaded deviating longitudinally. J China Coal Soc 36(10):1647–1652

Liu LM, Zhao SJ, Cao JZ, Zhang M, Zhu SG, Zhu TG (2015) D-type concrete-filled steel tube support in No.10 coal mine of pingdingshan. J Shandong Univ Sci Technol (Nat sci) 34(2): 7–13

Lu SL, Wang YH (1991) Ming induced influence on the roadways in weak surrounding rock and its controlling measures. J China Univ Min Technol 2(1):11–19

Luo Y, Chang JC (2009) Research on U-steel yieldable support with backfill technology in rock roadway. J Coal Sci Eng 15(4):355–360

Mitri HS, Hassani FP (1990) Structural characteristics of coal mine steel arch supports. Int J Rock Mech Min Sci Geomech 27(2):121–127

Qu GL (2013) Research on flexural performance of concrete-filled steel tubular support and its application. China University of Mining & Technology, Beijing

Rabcew LV (1960) New Austrian Tunnelling Method

Spearing S (2001) Shotcrete as an underground support material. Eng Fund Int Case Studies, Undergr Min Methods, pp 563–568

Sun LH, Yang BS, Sun CD, Li X, Wang ZW (2017) Experimental research on mechanism and controlling of floor heave in deep soft rock roadway. J Min Saf Eng 34(2):235–242

Tamada Kazuya, Klyoshi O, Kawamura A, Nishinura N (2006) Experimental study for estimating ultimate shear strength of U-shaped girders. Doboku Gakkai Ronbunshuu A 62(1):14–28

Tan XJ, Chen WZ, Liu HY (2017) A combined supporting system based on foamed concrete and U-shaped steel for underground coal mine roadways undergoing large deformations. Tunn Undergr Space Technol 68:196–210

Villaescusa E (1999) An Australian perspective to grouting for cable bolt reinforcement. Rock Support Reinf Pract Min:91–102

Wang LL, Gao WJ (2013) The perspective of 1000 m deep mining ecology of our country. N China Coal News 8:19

Wang Q, Li SC, Wang HP (2011) Mechanical property and economic effect of contractible concrete filled steel tube support. J Shandong University Eng Sci 41(5):103–113

Wang Q, Li WT, Li SC, Jiang B, Ruan GQ, Wang DC, Zhang SG, Liu WJ, Shao X (2015a) Field test study on mechanical properties of U-type confined concrete arch centering and support system in deep roadway. J Cent South Univ (Sci Technol) 46(6):2250–2260

Wang Q, Shao X, Li SC, Wang DC, Li WT, Wang FQ, Liu WJ, Zhang SG (2015b) Mechanical properties and failure mechanism of square type confined concrete arch centering. J China Coal Soc 40(4):922–930

Wang Q, Wang DC, Li WT, Ren YX, Ruan GQ, Xiao GQ, Guo NB, Pan R (2015c) Study on mechanical properties and deformation mechanism of U-type confined concrete arch centering. J China Coal Soc 40(5):1021–1029

Wang Q, Jiang B, Li SC, Wang DC, Wang FQ, Li WT, Ren YX, Guo NB, Shao X (2016) Experimental studies on the mechanical properties and deformation & failure mechanism of U-type confined concrete arch centering. Tunn Undergr Sp Tech 51:20–29

Wang Q, Jiang B, Shao X, Wang FQ, Li SC, Guo NB, Wang BQ, Xiao GQ, Pan R (2017) Mechanical properties of square steel confined concrete quantitative pressure-relief arch and its application in a deep mine. Int J Min Reclam Env 31(1):1–23

Wang Q, Jiang B, Pan R, Li SC, He MC, Sun HB, Qin Q, Yu HC, Luan YC (2018) Failure mechanism of surrounding rock with high stress and confined concrete support system. Int J Rock Mech Min Sci 102:89–100

Wang Q, Jiang B, Yang J (2019) Control Theory and Engineering Practice of Confined Concrete in Underground Engineering. Science Press, Beijing

Witthaus H, Polysos N, Witthaus H (2006) Applied geomechanics for support designin German deep coal mines. In: Proceedings of the 25th international conference on ground control in mining west virginia. Morgantown pp 199–208

Xie HP (2019) Research review of the state key research development program of China: deep rock mechanics and mining theory. J China Coal Soc 44(5):1283–1305

Xie WB, Jing SG, Wang T (2010) Structural stability of U-steel support and its control technology. Chinese J Rock Mech Eng 29(S2):3743–3748

You CA (2000a) Internal force calculation of U-supports considering yielding. Chinese J Geotech Eng 22(5):604–607

You CA (2000b) Calculation theory of roadway steel support. China coal industry publishing house, Beijing

Yu XF, Zheng YR, Liu HH (1983) Stability analysis of surrounding rock of underground engineering. Coal Industry Press, Beijing

Zang DS, Li AQ (2001) Study on concrete-filled steel tube supports. Chinese J Geotech Eng 23(3):342–344

Zhang DM, Liu ZS, Wang RL, Zhang DM (2019a) Influence of grouting on rehabilitation of an over-deformed operating shield tunnel lining in soft clay. Acta Geotech 14(4):1227–1247

Zhang N, Li XY, Zheng XG, Xue F (2013) Current situation and technical challenges of deep coal mining. China Symposium on mining technology of one kilometer deep coal mine Taian:10–31

Zhang QH, Yuan L, Yang K, Xue JH, Duan CR (2019b) Mechanism analysis on continuous stress-relief mining for preventing coal and rock dynamic disasters. J Min Saf Eng 36(1):80–86 + 102

Zhao YM, Liu N, Zheng XG, Zhang N (2015) Mechanical model for controlling floor heave in deep roadways with U-shaped steel closed support. Int J Min Sci Technol 25:713–720

Zheng YR (1988) Guide for design of bolting and shotcrete support for underground works. China Railway Publishing House, Beijing

Zhou XP, Huang YL, Ding ZH (2002) Formula of ultimate bearing capacity of Taisha base considering intermediate principal stress. J Stone Mech Eng 21(10):1554–1556

Chapter 2
Development of the High Strength Support System

We are often faced with challenging problems in deep excavations of underground engineering with soft surrounding rocks, such as large deformation, floor heave and failure of supporting components. The existing support methods do not meet the surrounding rock control requirements. So, it is very crucial to conduct research and development on new theories and technologies for surrounding rock control to guarantee the safety and the high efficiency of the deep excavation. In this chapter, a control concept of "high strength, pressure yielding and integrity" is put forward for surrounding rock control in the deep underground engineering with soft rocks; and a new high strength support system is developed, and its key technologies, design principles and construction methods are introduced.

2.1 The Concept of Surrounding Rock Control in Deep Underground Engineering

1. The control concept of "high strength"

Underground excavation is known to cause stress redistribution of the initial stress field. For excavation in soft rocks, the resultant stress state may exceed the strength of the rock, which leads to damage and reduced safety. The combined support of anchor, shotcrete, steel mesh and steel arch cannot meet the stability requirement of surrounding rock.

The traditional support method of anchor net spray is limited in improving the overall support strength of the deep underground engineering. Since the arch is the key support component, the increase of its strength will improve the overall support strength and thus control the deformation of the surrounding rock.

The high strength arch support is not in conflict with the flexible support of anchor net spray in New Austrian Tunnelling Method (NATM). NATM is not simply a flexible support, the essence of which is for fully mobilizing the bearing capacity of

Q. Wang et al., *High Strength Support for Soft Surrounding Rock in Deep Underground Engineering*, https://doi.org/10.1007/978-981-15-3844-5_2

the surrounding rock through supporting functions. With this method, the surrounding rock becomes the main structure to bear the upper pressure. The high strength arch support is not simply to provide rigid support. The essence of it is to provide a greater radial force around the surrounding rock and so that to form a high strength, stable load-bearing structure. Therefore, a new support system with high strength arches as the key could ensure the effective control of surrounding rocks in the deep underground engineering with soft rocks.

2. The control concept of "pressure yielding"

Since there is a great possibility of large deformation in the deep underground engineering, the high strengthened structure itself is not enough; the pressure should be well absorbed and transferred. The energy release of the surrounding rock maximize the self-supporting capacity for rock mass.

In addition, the surrounding rock deformation of underground engineering shows insurmountability to a certain extent. The load on the supporting structure sharply increases with the increase of the surrounding rock deformation; the structure is easy to reach its strength limit and then fail; as a result, the surrounding rock suffers further destruction. So, the support structure should have a pressure yielding function to release insurmountable deformation and reduce its own damage.

It is the key for the pressure yielding to maintain high support strength during a certain allowed deformation of surrounding rock; and therefore a high post-peak strength and its own bearing capacity of the rock is guaranteed. So, the pressure yielding here must be adjustable. That means the supporting system can release the pressure slowly and quantitatively rather than freely under the conditions of high strength support.

In the past decades of engineering practice, scientists and technicians have developed several supporting materials and structures that have pressure yielding functions such as: retractable U-steel pressure yielding, anchor bolt (cable) external pressure yielding ring, pressure yielding anchor bolt (cable) and excavation by steps with composite pressure yielding. At present, the quantitative pressure yielding of the anchor bolt (cable) can be realized. But as the core support component of the deep underground engineering, U-steels are cannot achieve quantitative pressure. It is necessary to research and develop a quantitative pressure yielding device for the supporting arches.

3. The control concept of "integrity"

The field survey of the soft rock underground engineering shows that the gap between the arch and the surrounding rock is the key factor to prevent the arch from fully playing its bearing capacity.

Our research shows that the supporting resistance of the soft rock underground engineering depends not only on the strength of the arch but also on the interaction between the arch and the surrounding rock. The insufficiency of the current excavation methods and supporting technologies inevitably leads to partial top falling or rib spalling and then to irregular cavities on the surrounding rock during the excavation. This leads to many irregular gaps between the surrounding rock and erected arches after the excavation. The contact between those irregular points and lines makes the arch subjected to concentrated loads or eccentric loads; and as a result, the arch suffers bending and torsional deformation, and loses its bearing capacity. The bearing capacity of the most suffered arch is only 6.4% of the arch under uniform load.

The wall-backfilling between the arch and the surrounding rock could solve the above problems. Filling materials is used to fill in the gaps behind the arch and form an integral bearing system consists of the surrounding rock, the filled object and the supporting structure; so that the pressure of the surrounding rock can be even on the arches; the self-bearing capacity of both the arch and the surrounding rock can be fully played; and therefore, the large deformation of the soft rock underground engineering can be controlled effectively.

Generally speaking, there are many problems with the existing supporting methods for the complex deep underground engineering with soft rocks. It is necessary to establish a new supporting system that can meet the support concept of "high strength, pressure yielding, and integrity".

2.2 The New Support System of High Strength

2.2.1 The Composition of the New Support System of High Strength

A new support system is developed, based on the concept of "high strength, pressure yielding and integrity". This system is shown in Fig. 2.1.

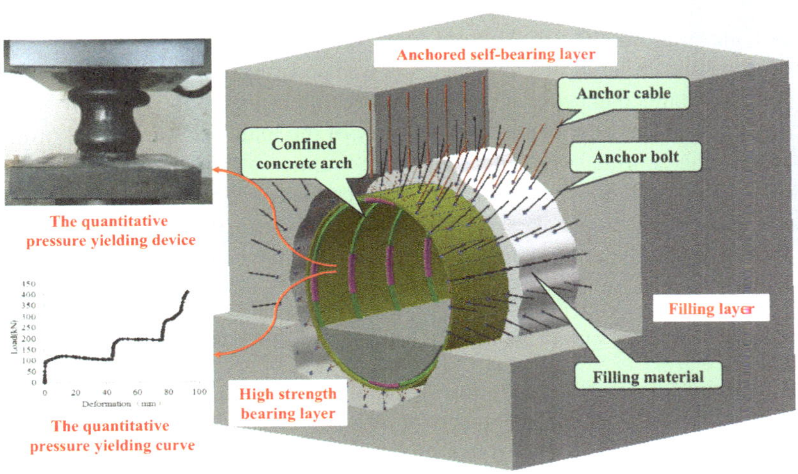

Fig. 2.1 Composition of new support system

The support system consists of three bearing layers:

(1) The internal high strength bearing layer: This layer is formed by the confined concrete arches with the function of quantitative pressure yielding. It is the main body for maintaining the integrity and effectiveness of the self-bearing of the surrounding rock.
(2) The external anchored self-bearing layer: This layer is formed by the surrounding rock reinforced by the anchor bolt (cable). It is the main body to bear the pressure of the surrounding rock.
(3) The intermediate adjustable filling layer: This layer is formed by filling flexible materials between the arches and the anchored surrounding rock. It is for achieving the integrity of the internal and external bearing structures.

2.2.2 The Principle of the New Support System of High Strength

The new support system is fully adapted to the control concept of "high strength, pressure yielding, and integrity" for the deep underground engineering with soft rocks. As a new type of high strength support, the confined concrete support is external confined materials filled with concrete.

The specific operating principles are:

1. The high strength support is mainly reflected in two aspects:

   (1) The arch can achieve the force symbiosis of the confining material and the internal concrete. In a three-direction compressive state due to the external constraints, the concrete core is much strengthened. Simultaneously, it provides a more effective support to the external constrained structure and prevents the structure from being depressed and destabilized. "The symbiosis" of the mechanical properties of the two fully exerts the strength of all the materials and improves the efficiency of the material mechanical properties. The confined concrete arch has 2–3 times of bearing capacity than the U-steel arch does with the same steel content.
   (2) The wall-backfilling makes the stress more uniform on the new high strength arch, reduces the stress concentration and guarantees the fully play of the arch bearing capacity.

2. The pressure yielding support is mainly reflected in three aspects:

   (1) The initial deformation and failure of the surrounding rock is insurmountable in the deep underground engineering with soft rocks. The anchor net spray method as the initial support can take advantage of its flexibility to realize the initial pressure yielding.

(2) The flexible wall-backfilling materials can achieve the pressure yielding on the intermediate adjustable layer of the filler through its own deformation.
(3) The quantitative pressure yielding devices are arranged on the secondary supporting arches with specific pressure yielding points and quantity. The devices are used to realize the post pressure yielding of the bearing structure and prevent the arch from being yielded and failed due to the excessive pressure. They could guarantee the integrity of the support system while releasing the deformation energy of the surrounding rocks.

3. The integral support is mainly reflected in two aspects:

(1) The anchor net spray, the high strength arch, and the wall-backfilling can ensure the integrity of the bearing system. Wall-backfilling can ensure the arch to contact in effective with the surrounding rock timely, withstands uniform load; it also can guarantee the high strength bearing capacity of the arch to exert and optimize the stress state of the underground engineering. The filling material connects the internal and external bearing structures together to form an integral body. The three bearing layers will join together to prevent partial failures of the support system effectively and improve the stability of the supported underground engineering.
(2) According to the main unstable parts of the surrounding rock and the first failed supporting components, targeted reinforcement is necessary to avoid the barrel effect and ensure the integrity of the support system in the bearing process. In the field practice of the deep underground engineering support, the failure of a single protective component often causes the failure of the entire support system. So, the bearing capacity of the support system is determined by the lowest-strength components in the support system. In the design and the field practice of the system, it is necessary to grasp the weaknesses and take effective measures to avoid or reinforce it.

2.3 The Key Technology of the New Support System of High Strength

2.3.1 The High Strength Confined Concrete Arch

The most common cross-section of the new high strength arch is square, circular or U-shape. And corresponding to its section form, the arch is named as the arch of square confined concrete (SQCC), circular confined concrete (CCC) or U-shaped confined concrete (UCC). It can be designed with different numbers of sections and shapes according to the actual needs of the field. Its main shape is semicircle with straight leg, circular, three-heart arch or horseshoe. The number of sections is 3–5.

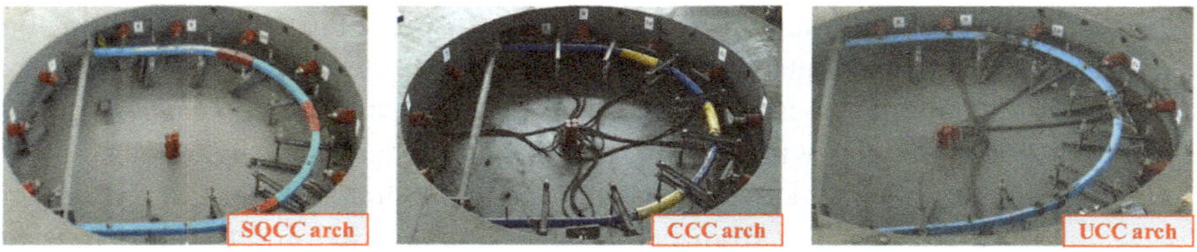

Fig. 2.2 The confined concrete arch

A whole arch is set up by each arch section connected by casings with grouting holes and vent holes arranged in the designated positions of the arches, as shown in Fig. 2.2.

2.3.2 The Quantitative Pressure Yielding Joints

The quantitative pressure yielding joint is a quantitative pressure yielding device installed in the casing joint. The quantitative pressure yielding of the arch can be achieved by the deformation of the device itself when the stress on the arch reaches a certain limit. The device has a specific pressure yielding point and quantity. It will be in many specific load-displacement relationship forms when being subjected to pressure. It will be in a constant resistance pressure yielding form when the deformation continues, but the load remains constant as the pressure reaches a certain level; it will be in a increasing resistance pressure yielding form when both the deformation and the load increase slowly; or it will be in a form of pressure yielding by stages. The quantitative pressure yielding joint is simple in the structure and easy to be installed. Its pressure yielding time and quantity can be controlled. So, it is suitable for supporting the arches in underground engineering, especially in all the difficult roadways, tunnels, chambers and others in the deep with soft rocks.

2.3.3 The Wall-Backfilling

The material used for the wall-backfilling must have a high bearing capacity and good compressive mechanical properties, such as the foamed concrete. This material is easy to find and saves costs with a good pumping ability. The pump is used to fill the material effectively in the space between the arch and the surrounding rock; so the arch can be subjected to a uniform force; the high strength bearing capacity of the arch can be fully exerted; and the labor intensity of the worker can be reduced. In addition, the pressure yielding can be realized on the intermediate adjustment layer by the deformation of the filling material itself as the arch is under the pressure.

2.3.4 The Support System Design and Construction Methods

1. The Support System Design

The main aspects of the new support system design includes: the actual excavation dimension of the underground engineering, the shape and size of the arch, the section numbers of the arch, the parameters of the arch cross-section, the parameters for the pressure yielding joints (points, quantity etc.), the location and the size of the grouting holes and the vents, and the strength and the matching ratio of the concrete core (Fig. 2.3).

Pay attention to the following points in the specific operation:

(1) The actual excavation dimensions of the underground engineering must meet the requirements for the arch installation and the implementation of the wall-backfilling.
(2) In general, the arch is in a shape of semicircle with straight leg, circular, or horseshoe. The arch in a shape of semicircle with straight leg is suitable for the underground engineering with a large pressure on the vault; the circular arch is suitable for the underground engineering with a large overall pressure; and the horseshoe-shaped arch is similar to the circular arch and its cross-section utilization is higher.
(3) The positions of the grouting and vent holes are designed according to the requirement of grouting upward from the bottom. Each section of the arch needs to be equipped with a pair of grouting and vent holes; the grouting hole should have its diameter not less than 70 mm and its position as downward as possible after the other requirements are met; and with an appropriate size of 30 mm, the vent hole should be set as upward as possible after the other requirements are met.
(4) The self-compacting concrete is adopted generally for the concrete core inside the arch to ensure the strength of the arch. The concrete core should have sufficient fluidity.

2. The Confined Concrete Arch and its Auxiliary Components Processing

The main elements are: the left arch leg, the right arch leg, the left arch string, the right arch string, the casing, the tie rod, the tie rod ring, the metal mesh plate, the skewback bearing plate and the others.

3. The Construction Process

The construction sequence: Assemble upward from the bottom and the construction process is shown in Fig. 2.4. Install the left and right arch legs in the design position → install the pressure yielding device at the connecting cross-section of the arch leg and the arch string → install the left and right arch strings → set the guard plate near the springing point and the skewback of the arch and pass the anchor bolt through the guard plate to fix it in the rock → the arches are connected by the tie rod which are installed respectively at the spandrel and the straight leg.

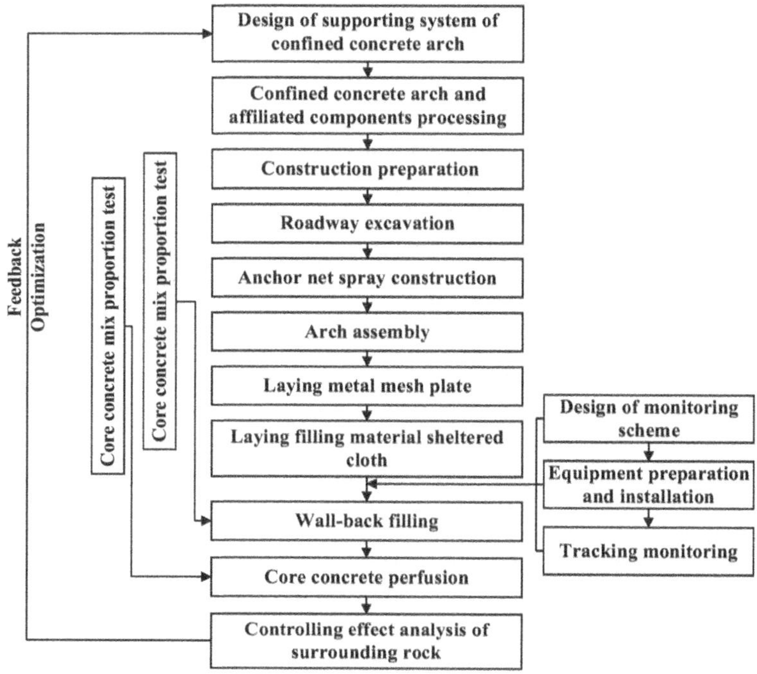

Fig. 2.3 Design and construction process

4. The Construction Principles

The uniaxial compressive strength of the wall-backfilling material should be above 5 MPa, and so, the material has good toughness and deformation resistance; the early strength of the material should reach at least 2 MPa within 24 h; and the filling material should be compacted with no obvious void left.

With micro-expansion, the fine stone concrete is adopted as the concrete core to be perfused into the arches with pumps. The early strength of the material is higher. The lifting-up technology is adopted to grouting upward from the bottom. The grouting hole is lower, and the vent is upper. The arch leg is perfused first and then the arch string follows. The perfusion should be plump with the concrete overflowing from the vent.

2.4 Chapter Summary

A brief description of this chapter is as follows:

(1) This chapter proposes the support concept of "high strength, pressure yielding, and integrity"; and the rigid and flexible combination support methods to

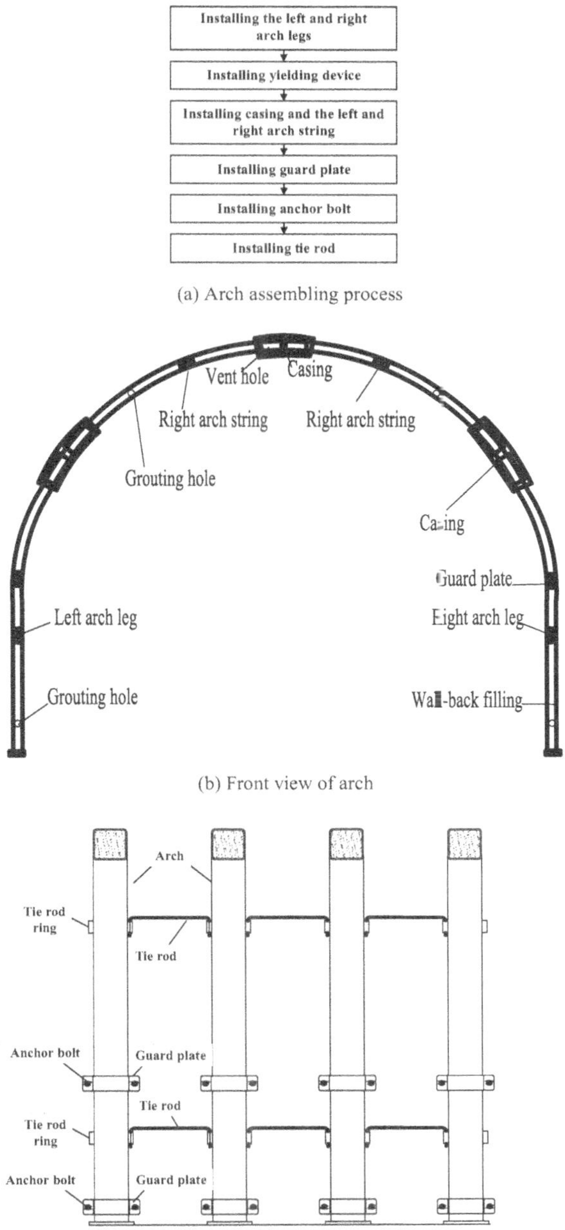

(a) Arch assembling process

(b) Front view of arch

(c) Side view of arch

Fig. 2.4 Arch assembling process and assembly effect

ensure the effective control of the surrounding rock of the deep underground engineering with soft rocks.

(2) The new high strength support system includes: The internal high strength bearing layer which is formed by the confined concrete arches with the function of quantitative pressure yielding. The external anchored self-supporting layer which is formed by the surrounding rock reinforced by the anchor bolt (cable). The intermediate adjustable layer of the filler which is formed by filling flexible materials between the arches and the anchored surrounding rock.

(3) The forms of the high strength arch include the square confined concrete (SQCC), the circular confined concrete (CCC) and the U-shaped confined concrete (UCC). The quantitative pressure yielding device has specific pressure yielding points and quantity; and it realizes the high-resistance quantitative pressure yielding of the bearing structure and ensures the integrity of the support system while releases the deformation energy of the surrounding rock.

Chapter 3
Mechanical Properties Test on the Basic Components of New High Strength Arches

In order to study the mechanical properties of the basic components of confined concrete proposed, which is the new high strength support, the laboratory and numerical tests of axial compression is carried out on the components in this chapter. The analysis is made on the deformation and failure mechanism and bearing mechanism of the basic component, and the mechanical properties of the components are studied as well. Research is carried out on the deformation of the basic components with grouting hole, and the reasonable reinforcement scheme is obtained by comparison and selection. Through the series of tests on the casing joints, the effect of different factors is clarified on the mechanical properties of the joints.

3.1 Axial Bearing Mechanism Test of the Basic Component

3.1.1 Test Scheme

The analysis is made through the axial compression and numerical tests on the SQCC and the traditional U-steel short columns. The deformation and failure mechanism and bearing mechanism are analyzed from the aspects of the deformation and failure mode, the load-strain curve, the ultimate bearing capacity, and the acoustic emission response characteristics of the concrete core.

The short column types are shown in Table 3.1 for the SQCC and U-steel tests. The height L of the SQCC specimen is designed to be 3 times the length B of the square steel pipe, that is $L/B = 3$; the height L of the U-steel short column specimen is designed to be 3 times the cross-section length B (Fig. 3.1f as an example); and pressure bearing plates are welded to both the upper and the lower cross-sections of the specimen. The specific processing dimensions of the different columns in each group are shown in Fig. 3.1.

Q. Wang et al., *High Strength Support for Soft Surrounding Rock in Deep Underground Engineering*, https://doi.org/10.1007/978-981-15-3844-5_3

Table 3.1 Test schemes

Serial number	Types	Specimen types
1	SQCC short column	SQCC140*6
2		SQCC140*8
3		SQCC150*8
4	U-steel short column	U25
5		U29
6		U36

Note Taking "SQCC140*6" as an example, it represents a short column specimen of square confined concrete. The cross-section edge length is 140 mm and the steel pipe wall thickness is 6 mm

Three sets of loading tests are made on the same specimen of each type. The test results are summarized in Table 3.1.

3.1.1.1 Laboratory Loading Scheme

The laboratory axial compression test of the short column is performed on a 1000 t press, as shown in Fig. 3.2. The short column specimen is geometrically aligned with the pressure loading device during the test. Two displacement meters are set along the longitudinal direction of the specimen to measure the longitudinal deformation. An acoustic emission instrument is used to analyze the failure of the concrete core under the axial compression of the confined concrete short column.

The step loading method is adopted in the tests. Before reaching 50% of the predicted ultimate load N_{up}, the load of each stage is 1/10 of N_{up}, and the pressure holding time of each stage is about 2 min; when reaching 50–80% of the predicted ultimate load N_{up}, the load of each stage is 1/15 of N_{up}, and the pressure holding time of each stage is about 2 min; when the load reaches 80% of N_{up}, the specimen is loaded continuously at the slow speed until damaged.

3.1.1.2 Numerical Test Scheme

Based on the laboratory tests, the numerical tests are carried out on the axial compression of the short column; the numerical test results are compared with the laboratory test results to verify the rationality of the numerical models and material parameters.

1. Material Constitutive Relations

(1) Steel Constitutive Relation

The stress-strain (σ-ε) curve of the steel of SQCC short column is shown in Fig. 3.3. Its yield strength f_{y-sc} is 409 MPa, its ultimate strength f_{y-sc} is 594 MPa, and its elastic modulus E_{s-sc} is 204 GPa.

The stress-strain (σ-ε) curve of the steel of U-steel short column is shown in Fig. 3.4. Its expression is:

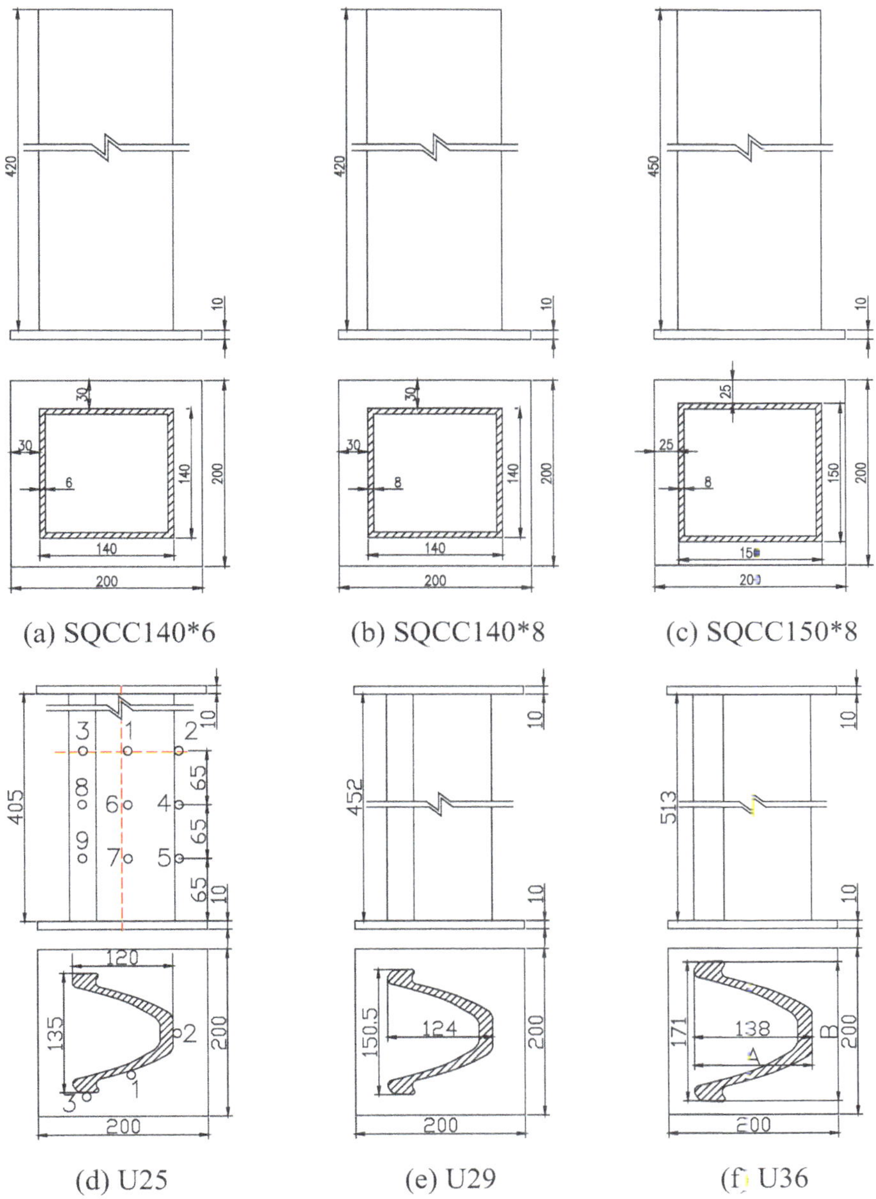

Fig. 3.1 Short column size (mm)

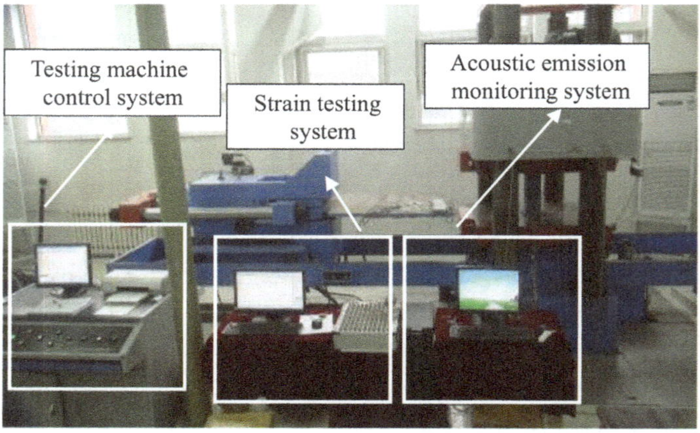

Fig. 3.2 Loading and monitoring system

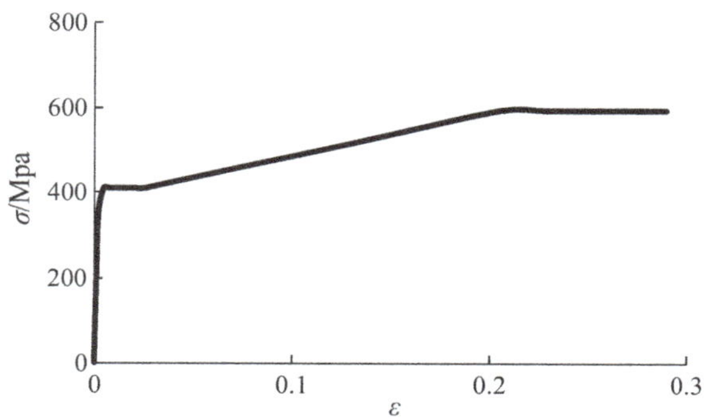

Fig. 3.3 Weighted σ-ε curve

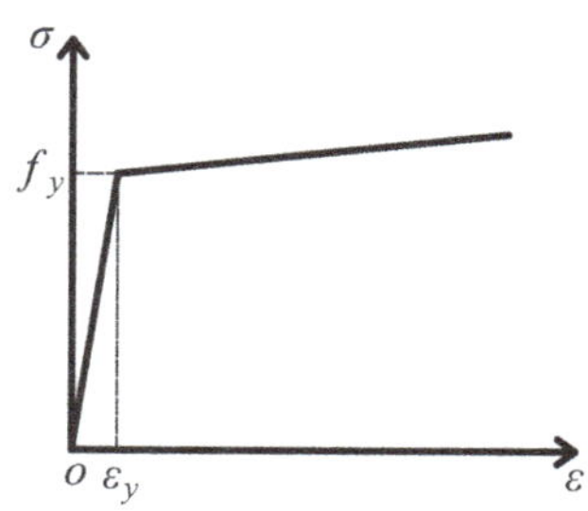

Fig. 3.4 Bilinear model

$$\sigma = \begin{cases} E_{s-u}\varepsilon & (\sigma \le f_y) \\ 0.01E_{s-u}(\varepsilon - \varepsilon_{y-u}) + f_{y-u} & (\sigma > f_y) \end{cases} \quad (3.1)$$

In the formula, the elastic modulus of the steel E_{s-u} is 197 GPa; the elastic modulus of the reinforced section is 0.01 E_{s-u}; the yield strength of the steel f_{y-u} is 335 MPa, and the ultimate strength f_{u-u} is 533 MPa.

(2) Concrete Constitutive Relation
Considering the influence of the confinable effect coefficient of the steel pipe on the concrete core, the plastic damage model is adopted for the concrete core (Han 2019), and its expression is:

$$y = \begin{cases} 2x - x^2 & (x \le 1) \\ \frac{x}{\beta_0(x-1)^\eta + x} & (x > 1) \end{cases} \quad (3.2)$$

In the formula: $x = \frac{\varepsilon}{\varepsilon_0}$; $y = \frac{\sigma}{\sigma_0}$; $\varepsilon_0 = \varepsilon_c + 800\xi^{0.2} \times 10^{-6}$; $\varepsilon_c = (1300 + 12.5f_c') \times 10^{-6}$; $\eta = 1.6 + 1.5/x$;$\beta_0 = \frac{(f_c')^{0.1}}{1.2\sqrt{1+\xi}}$; and the physical meaning of the symbol ε, ε_0, ε_c, σ_0, β_0, η, $f_c{}'$, ζ can refer to the reference (Han 2019). The compressive strength of the concrete in the above formulas is in MPa.

2. Elements Setting

The three-dimensional solid elements are adopted for both the steel pipe and the concrete core; meanwhile, with the consideration of both the calculation accuracy and cost, the hexahedral element in the reduced-integral format is adopted for both the steel pipe and the concrete core; and the element type is selected as C3D8R or C3D20R.

3. Interaction

Tie is used for the restraint between the inner wall of the steel pipe and the concrete core.

3.1.2 The Result Analysis

3.1.2.1 Deformation and Failure Mechanism

The observation on the entire process of specimen loading indicates that U-steel and SQCC short columns have remarkably different failure modes after losing stability under compression. Figures 3.5 and 3.6 show typical failure modes of some specimens.

The comparative analysis is as follows:

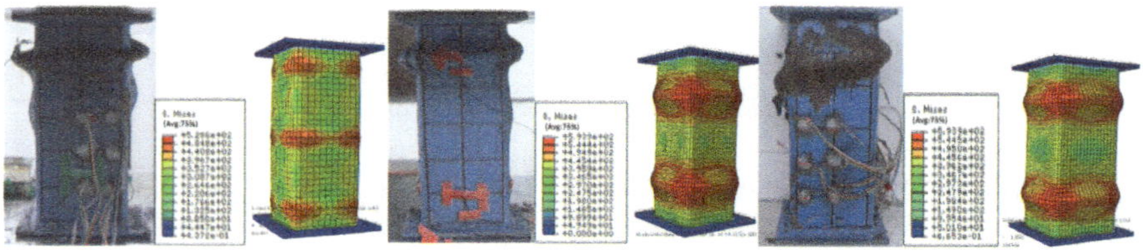

(a) SQCC140*6 (b) SQCC140*8 (c) SQCC150*8

Fig. 3.5 SQCC short column failure modes

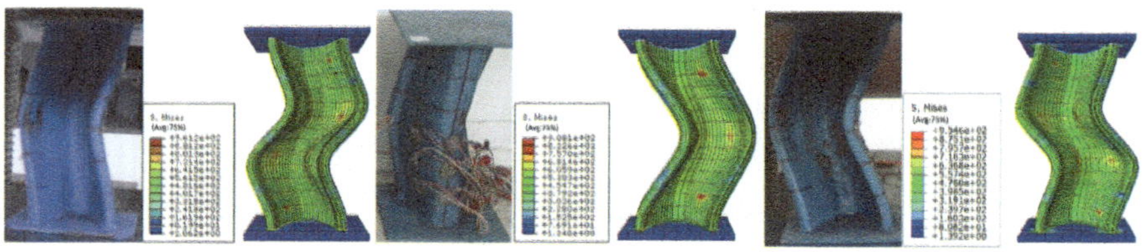

(a) U25 (b) U29 (c) U36

Fig. 3.6 U-steel short column failure modes

(1) SQCC short columns have shear slip and drumming failure modes. In the early loading stage, no obvious deformation failure occurs on SQCC short columns. Along with the load increase, the specimens enter the elastic-plastic failure stage; shear slip lines have appeared on the steel pipe wall. When the load approximates 70% of the ultimate load, obvious buckling wave peaks occur on the specimen surface; meanwhile, more shear slip lines appear and gradually cover the pipe wall. As the load continues to increase, the buckling wave becomes more obvious and develops from the original half-wave of the only one peak to 1.5–2.5 complete buckling waves.
(2) The U-steel short columns have a failure mode of bending instability. The specimens are in elastic deformation state at the initial stage of loading, and there is no obvious deformation and failure on the specimens. With the increase of load, the specimens quickly reach their ultimate bearing capacity and then become buckling after entering the yield stage, and their bearing capacity is significantly reduced. The Figs. 3.6a, b, c show that the U-steel short columns have a significant bending instability under the axial compression with the most serious deformation on the ears of their cross-sections. This shows the overall stability of the specimens is poor, and the bearing capacity of the specimens after yielding mainly depends on the flexural rigidity. The bearing capacity has not been effectively exerted. The above-mentioned destruction and instability mode is the main reason why U-steel arches often generate out-plane instability damage in complex condition roadways.

The above analysis shows that the SQCC short columns have better ductility and post bearing capacity under axial compression. The specimens have an overall plastic instability and eventually shows a shear slip or drumming failure mode. This feature effectively avoids the bending instability of U-steel short columns and helps to exert its post bearing capacity.

3.1.2.2 Bearing Mechanism

Figure 3.7 shows the axial force (N) and the average longitudinal strain (ε) relationship curves of the specimens. Taking Fig. 3.7a as an example, N in U25-N represents the axial force of the numerical test and L in U25-L represents the axial force of the laboratory test. According to the test results, the typical N–ε curves of SQCC short columns and U-steel short columns are made and shown in Fig. 3.7d.

The analysis shows:

(1) $N-\varepsilon$ relationship curve of the U-steel short column generally consists of three stages: elastic deformation (OA′)—elastic-plastic deformation (A′B′)—bearing capacity descent (B′C′). In OA′ stage, the steel is basically in the elastic state, and its deformation is not large, but the load increases fast. The point A′ roughly corresponds to the beginning of the steel entering the elastic-plastic stage, and

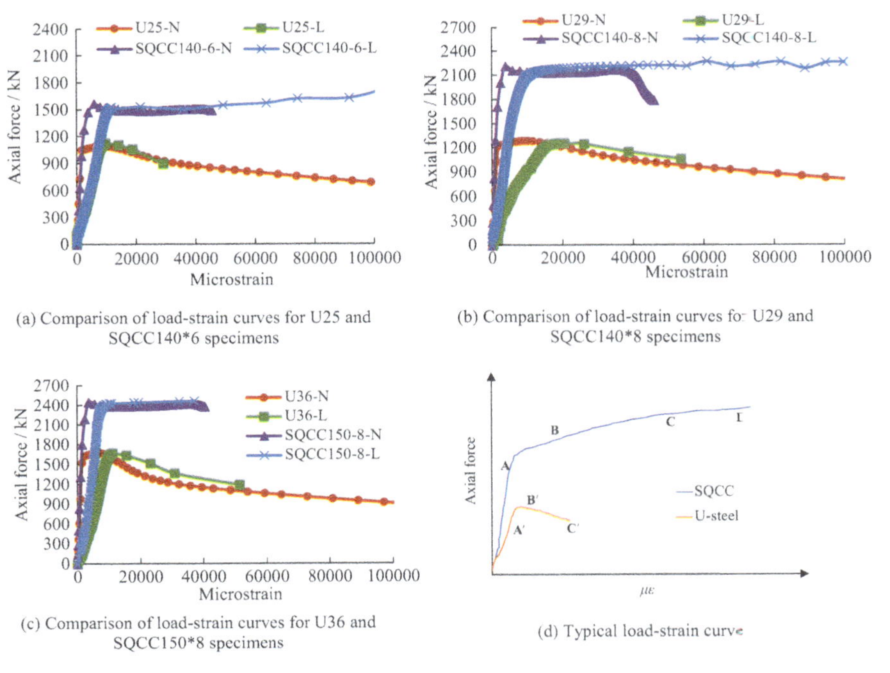

(a) Comparison of load-strain curves for U25 and SQCC140*6 specimens

(b) Comparison of load-strain curves for U29 and SQCC140*8 specimens

(c) Comparison of load-strain curves for U36 and SQCC150*8 specimens

(d) Typical load-strain curve

Fig. 3.7 Load-longitudinal average strain curve

the corresponding load value is 92.2% of the ultimate load N_{ue}. During the A′B′ stage, the steel is in the elastic-plastic state. When it reaches the point B′, the short column enters the critical point of bending instability. The corresponding load value is 99.4% of the ultimate load N_{ue}. The curve in the stage B′C′ starts to decline after reaching the peak point B′, and significant bending instability occurs on the steel pipe at point C′. At this stage, the load decreases significantly with the development of deformation.

(2) The $N-\varepsilon$ relationship curve of the SQCC short column generally consists of four stages: the elastic deformation (OA), the elastic-plastic deformation (AB), the plastic deformation (BC) and the rapid deformation (CD). The $N-\varepsilon$ relationship curve does not show the decreasing trend of the U-steel short column after reaching point B. Instead, it shows a continuously and stably increasing trend, but the speed obviously slows down. Finally, the curve turns into an approximate straight line and the load stops to increase.

The above analysis shows that there are three stages in the $N-\varepsilon$ relationship curve of the U-steel short column from rise to fall. There are four stages in the curve of the SQCC short column from rapid rise to gentle rise, without fall stage. The SQCC short columns have a better ductility and post bearing capacity.

3.1.2.3 Fracture Mechanism of the Concrete Core

Figure 3.8 shows the acoustic emission test (AE) results and the relationship diagram corresponding to the typical SQCC load-axial average strain. According to the analysis results of the previous test phenomenon, the short column failure mode, the $N-\varepsilon$ curve, and the ultimate bearing capacity, a comprehensive analysis is made on the deformation and failure mechanism of the SQCC short column under axial compression.

In the whole process of loading until destruction occurs, acoustic emission activities experience four periods:

(1) Incubation period: In the early loading stage, the confined concrete short column is in the elastic deformation stage. Both AE count and energy are small. The internal crack of the concrete is in the incubation state and relatively inactive. The analysis results on the longitudinal strain measurement of the steel pipe show that the load on concrete core is lower in this stage than that in the later loading stage. In addition, the concrete bears the three-dimensional pressure under the confinement effect of the steel pipe. As a result, crack and expansion are restricted and some cracks are even closed under the compressive stress.

(2) Active period: As the load increases, the confined concrete short column enters the elastic-plastic deformation stage (65–85% of the ultimate load). In this stage, both AE count and energy obviously increase. This indicates the steel pipe fails to effectively control the transverse deformation increase of the concrete. Larger cracks appear on the concrete. Moreover, with the further increase of load, a large number of micro-cracks converge and some become penetrating

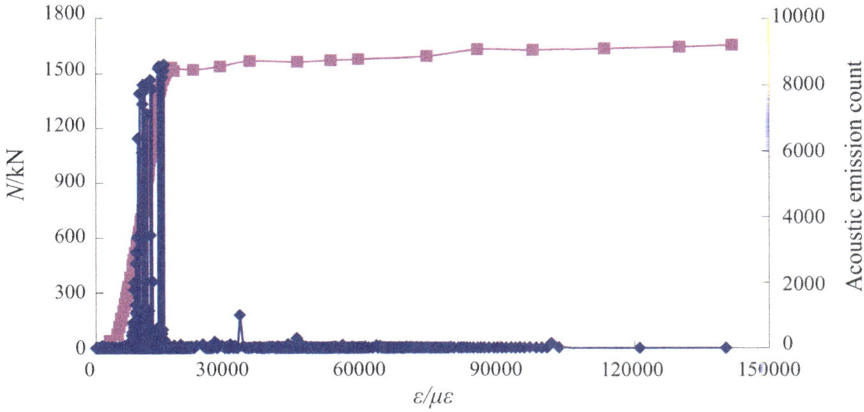

(a) Acoustic emission count

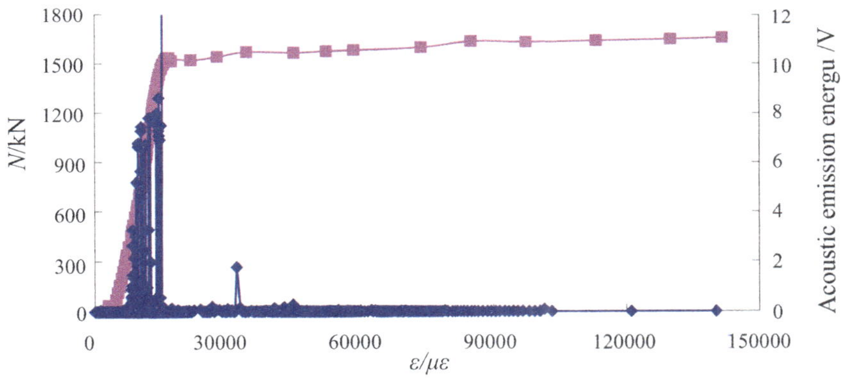

(b) Acoustic emission energy

Fig. 3.8 Acoustic emission test results

cracks. Meanwhile, enough elastic energy is accumulated inside the concrete core. During the energy releasing, the deformation begins to accelerate, and the load increases slowly. Obvious shear slip lines and drumming failure appear on the short column surface. Therefore, the point B on the $N–\varepsilon$ curve of the SQCC short column could be regarded as the sign to judge the overall failure of the concrete core.

(3) Stabilization period: With the increasing load, the confined concrete short column enters the plastic deformation stage. In this stage, both AE count and energy obviously decrease. This shows the penetrating cracks inside the concrete core become more and denser after the confined concrete pASRes the point B on the $N-\varepsilon$ curve. Moreover, the original concrete core basically has changed from

a complete block to accumulated fragment, and less energy is released in the fragmented block extrusion and mechanical interlocking process. In this stage, the load increased slowly. It could be considered that even though the steel pipe has entered the complete plastic failure state, the load increment is borne by the concrete core. The increase of concrete core bearing capacity exceeds the decrease of the steel pipe bearing capacity. Therefore, the specimen's load-strain curve does not have a decrease stage.

(4) Silence period: Finally, the confined concrete short column enters the failure stage. In this stage, the AE activity disappears basically, and the specimen shows the shear slip or drumming failures. Due to the external confined of the steel pipe, the entire confined concrete still has bearing capacity. The load stops to increase, and the entire deformation continues. From the perspective of failure modes, this belongs to complete failure state.

The above analysis shows the AE activity well correlates with the failure process of the specimens. Therefore, the analysis on AE characteristic parameters could provide an effective method for studying the concrete core fracture mechanism and the performance optimization of coupling with the confining steel pipe.

3.1.3 Summary

(1) The SQCC short column presents overall plastic instability, which avoids the bending instability as the traditional U-steel short columns occurred; the typical $N-\varepsilon$ curve of the SQCC short columns shows four stages from rapid rise to gentle rise without fall stage. It has better ductility and post bearing capacity.

(2) Acoustic emission activity of the SQCC short column has experienced four periods: incubation → active → stabilization → silence, corresponding with the four stages of the $N-\varepsilon$ curve. When the load is 68–85% of the ultimate load, the acoustic emission activity of the specimen is in the active period. In this period, larger cracks begin to appear in the concrete core, and penetrating cracks are formed. The point B of the $N-\varepsilon$ curve could be regarded as the sign to judge the overall failure of the concrete core.

3.2 Component Reinforcement with Grouting Holes

In the construction with the confined concrete support system, grouting holes need to be arranged on the arch for the concrete to be grouted. The grouting holes become key parts of the arch with local weakening and stress concentration effects and greatly influence the bearing capacity of the whole arch. Therefore, it is necessary to reinforce the grouting holes in order to ensure the overall strength of the arches.

In this section, axial compression tests are carried out on the specimens under three different reinforcement schemes for the grouting holes. According to the numerical tests, comparative analysis is made on the key damaged parts the deformation and failure modes, the bearing mechanisms and ultimate bearing capacity of short columns under different reinforcement schemes. Finally, the reasonable reinforcement scheme is obtained through comparison and selection.

3.2.1 Influence Mechanism of the Grouting Hole

Two types of test specimens are designed for the axial compression tests of the short columns. The first type is the ordinary SQCC short column, and the second type is the SQCC short column with grouting holes (GH). "SQCC150*8" and "SQCC150*8-GH" present the edge length of the specimen is 150 mm and the thickness of the steel pipe is 8 mm of the SQCC short column with and without grouting holes, respectively. The height of the specimen is designed to be 3 times the length of the steel pipe as 450 mm. The diameter of the grouting hole is 80 mm and the distance between the hole center and the bottom of the short column is 225 mm. The specimen parameters of the short columns are shown in Fig. 3.9. Through the short column axial compression test, the deformation and failure modes and mechanical properties of the short columns are compared and analyzed. The weakening effect of the grouting hole is clearly clarified on the specimen.

1. Deformation and Failure Mechanism

The typical failure modes of some specimens are shown in Fig. 3.10.

The deformation and failure modes of the two short columns are compared and analyzed after the columns are compressed and instable.

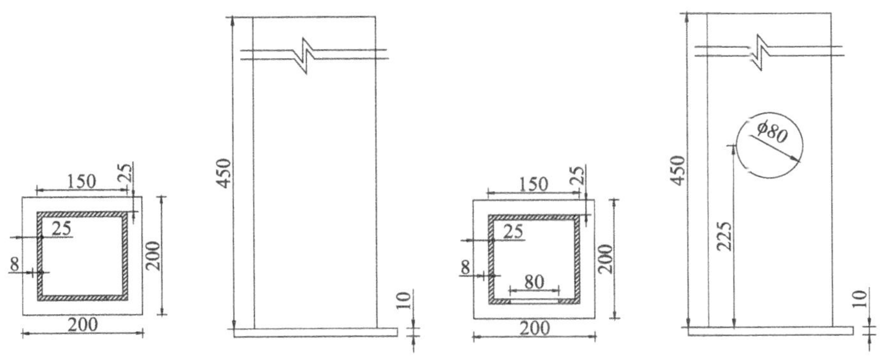

(a) SQCC150 * 8 (b) SQCC150 * 8-GH

Fig. 3.9 Short column dimensions (mm)

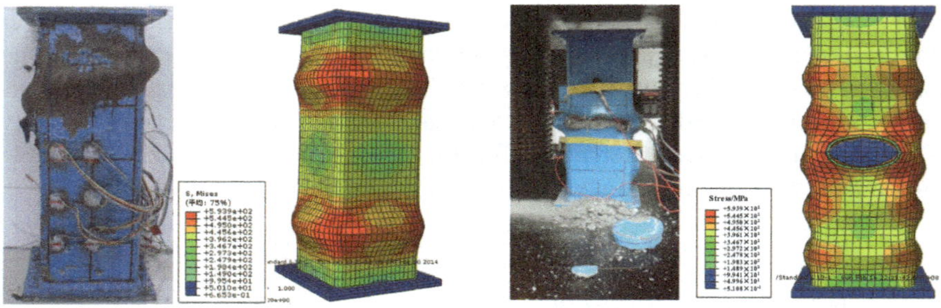

(a) SQCC150 * 8 (b) SQCC150 * 8-GH

Fig. 3.10 Typical failure modes of specimens

(1) The ordinary SQCC short columns exhibit a drumming failure mode. The short column specimens have no obvious deformation during initial loading. As the load increases, shear slip lines begin to appear on the local steel pipe wall. As the load continues to increase to about 70% of the ultimate load, obvious buckling waves appear on the surface of the specimen. As the load further increases, the short columns are deformed and damaged.
(2) After the grouting holes are arranged, a clear and multi-folded drumming failure mode occurs on the short column. During the test, with the increase of load, the stress concentration begin to appear at the grouting hole and its nearby positions, and the grouting hole is damaged first. As the load continues, a significant buckling wave peak appears on the short column and instability failure is followed.

2. Bearing Mechanism

Figure 3.11 is the load-displacement comparison curves between the laboratory test and the numerical test. Figure 3.11a is the load-displacement comparison curves of the short column of SQCC without grouting hole, and Fig. 3.11b is the load-displacement comparison curve of the SQCC short column with grouting hole.

The analysis shows:

(1) The laboratory and numerical test curves of SQCC short columns with and without holes are in good consistency, and the load-displacement curves of different short columns have similar trends. They all experience the four stages of elastic, elastic-plastic, plastic, and failure.
(2) Compared with the SQCC short column without hole, the yield strength and the ultimate bearing capacity of the short column with hole have been reduced to a greater extent.

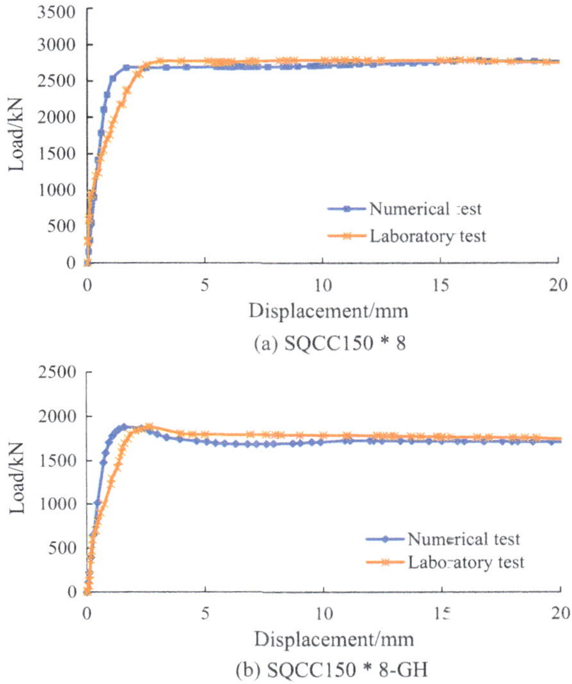

Fig. 3.11 Load and displacement comparison curves for laboratory and numerical tests

3. Comparative Analysis of Bearing Capacity.

Table 3.2 shows the comparison of the ultimate bearing capacity of the two types of short columns in both the laboratory and the numerical tests. This analysis shows:

(1) Comparing the results of the numerical and the laboratory tests on the two types of short columns, the maximum difference rate in ultimate bearing capacity is only 2.1%. The test results are basically consistent, which proves the rationality of the models, the material parameters and the load conditions for numerical tests.
(2) Compared with SQCC short column without hole, the axial bearing capacity of the short column with hole is reduced by 29.9% (according to the laboratory test results). This indicates that the axial compressive strength of the SQCC

Table 3.2 Comparison of laboratory and numerical test results

Short column type	Laboratory test ultimate bearing capacity/kN	Numerical test ultimate bearing capacity/kN	Difference rate/%
SQCC	2726	2685.5	1.5
SQCC-GH	1911.2	1871.1	2.1

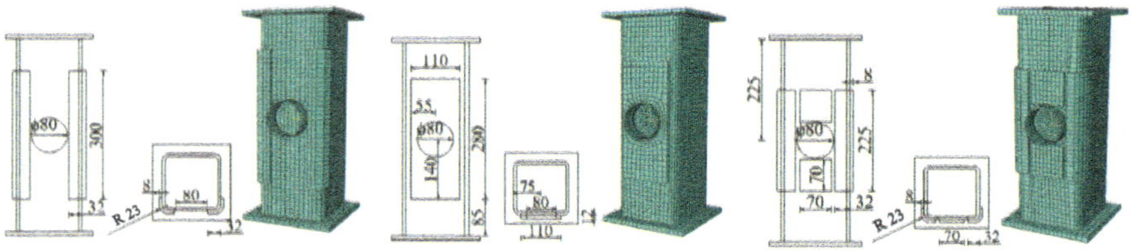

(a) ASR-Angle Steel Reinforcement (b) PPR-Perforated Plate Reinforcement (c) SPR-Surrounding Plate Reinforcement

Fig. 3.12 Dimensions and model diagrams for different reinforcement schemes (mm)

short column is significantly reduced with the hole. Therefore, it is necessary to reinforce the grouting holes in order to ensure the overall strength of the arches.

3.2.2 Reinforcement Scheme

To study the effect of different reinforcement schemes on the axial bearing capacity of SQCC with holes, some changes are made on the shape and the thickness of the reinforcement steel plate. Three numerical schemes are designed. They are the ASR-Angle Steel Reinforcement, PPR-Perforated Plate Reinforcement and SPR-Surrounding Plate Reinforcement.

In order to facilitate the comparative study on the short column reinforcement test, the designed short column parameters are invariants. The reinforcement steel plates are designed with the same steel content. The sizes of the ASR-Angle steel plate, PPR-Perforated plate and SPR-Surrounding plate are 306.5 cm^3, 309.3 cm^3, and 308.2 cm^3, respectively. The maximum difference rate is 0.9% among the three. The specific parameters are shown in Fig. 3.12.

3.2.3 Results Analysis

1. Deformation and Failure Mechanism

Figure 3.13 shows the failure modes of the SQCC short columns with grouting holes in numerical test with the three reinforcement schemes.

The deformation process and failure modes of the different short columns are compared and analyzed as follows:

(1) In the scheme of ASR, no obvious deformation and failure characteristics of the short column are observed in the initial stage of the test. Along with the increase of the load, the short column gradually enters the elastic-plastic deformation

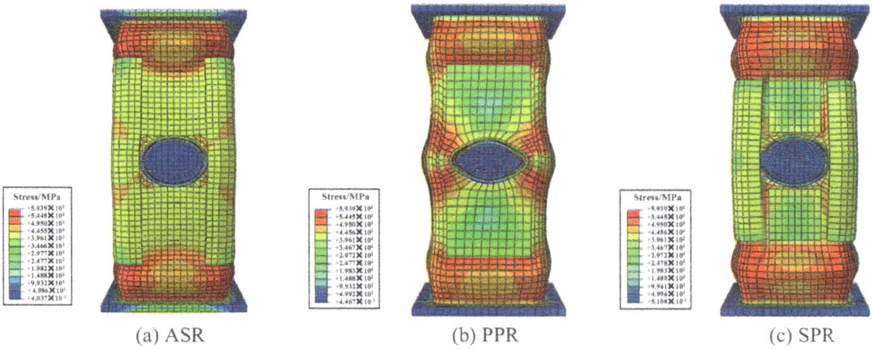

Fig. 3.13 Typical failure modes of short columns under different reinforcement schemes

stage. The shear slip lines appear above and below the reinforcement steel plate, and the short column deformation is small. As the load increases to about 80% of the ultimate load, buckling wave peaks appear above and below the steel plat. The buckling wave peaks become more obvious as the load increases. Throughout the test process, no significant deformation occurs at or around the grouting hole. The positions above and below the reinforcement steel plate become the key damaged parts, and the short column reinforcement effect is particularly significant.

(2) In the scheme of PPR, the short column has no obvious deformation and failure characteristics at the initial stage of the test. As the load increases, the stress concentration begins to appear at and around the grouting hole, and the deformation is small. As the load gradually increases to about 70% of the ultimate load, the grouting hole is the first to be flattened and damaged. It is the key failure position on the short column. As the load increases, the short column gradually loses its stability. Throughout the test process, the grouting hole is still a key damaged part, and the reinforcement effect is poor.

(3) In the schemes of SPR and ASR, the short columns have the same deformation characteristics at the initial stage of the test. They are in the elastic-plastic deformation stage with a smaller deformation. As the load increases to about 75% of the ultimate load, buckling wave peaks appear gradually above and below the plate. During the entire test process, there is no obvious deformation at and around the grouting hole. The key damaged part moves to the positions above and below the steel plate, and the reinforcement effect is obvious.

2. Bearing Mechanism

Figure 3.14 is the load-displacement curves of the reinforced SQCC short columns. Figure 3.14a is the load-displacement comparison curves of the short columns in the laboratory and the numerical tests under the ASR scheme. Figure 3.14b is the

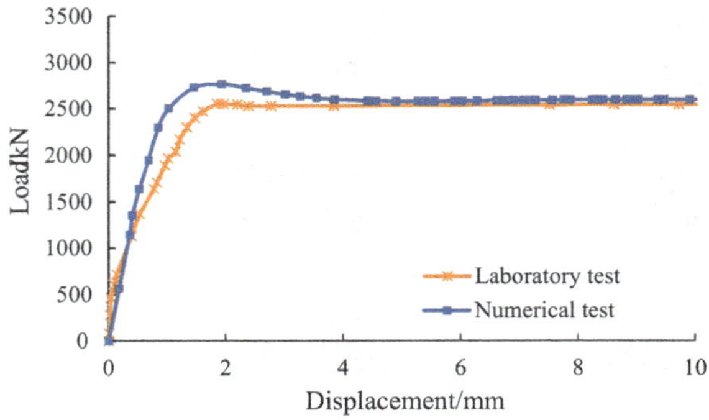

(a) Comparison of the reinforced short columns with ASR scheme in the numerical and laboratory tests

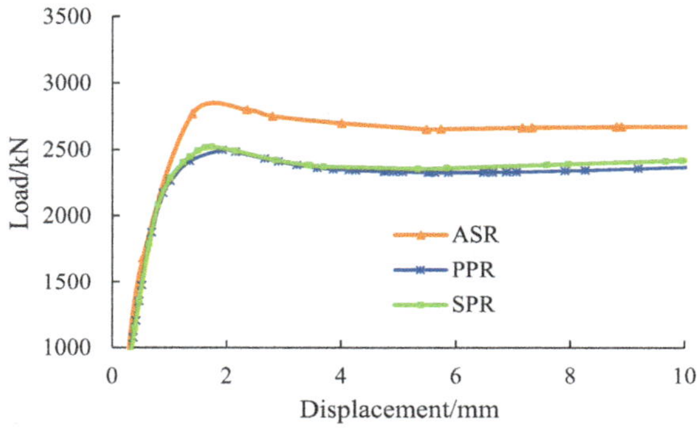

(b) Comparison of the reinforcement schemes

Fig. 3.14 Load-displacement curves under the different reinforcement schemes

load-displacement comparison curves of the short columns under the three schemes of ASR, PPR and SPR.

The following conclusions can be drawn from Fig. 3.14:

(1) The results of laboratory and numerical tests have a good consistency in the deformation characteristics and the load-displacement curves under the ASR scheme. This verifies the rationality of the numerical tests.

(2) The load-displacement curves of the different reinforced short columns with grouting holes are in similar basic shape. The curves can be divided into four stages: elastic deformation stage, elastic-plastic deformation stage, plastic deformation stage, and rapid deformation stage.
(3) When the load-displacement curves are in the elastic deformation stage, their linear slopes remain basically the same. This indicates the effect of the three reinforcement schemes is small on the structural rigidity and can be neglected when the designed steel content is the same for all the columns.

Table 3.3 Short column ultimate bearing capacity results

Reinforcement scheme	Ultimate bearing capacity/kN	Strength index α/%
Angle steel reinforcement	2782.48	148.7
Perforated plate reinforcement	2435.91	130.2
Surrounding plate reinforcement	2512.4	134.3

3. Strength Index Analysis

Table 3.3 is the results comparison and the analysis of the ultimate bearing capacity of the short columns in the different reinforcement schemes. The defined strength index α is the ratio of the ultimate bearing capacity of the short column before and after the reinforcement is made on the grouting holes.

Table 3.3 shows:

The strength indexes of the short column under the three reinforcement schemes of ASR, PPR and SPR are respectively 148.7%, 130.2% and 134.3%. This shows the ASR scheme has the best reinforcement effect with the same steel content. The ASR short column has the largest bearing capacity, and the stress concentration decreases most obviously. Therefore, it is more appropriate to reinforce the SQCC150*8 short columns with grouting holes by the ASR scheme.

3.2.4 Summary

After the SQCC short column is arranged with the grouting hole, the flattening failure occurs first at the position of the grouting hole under the axial compression. The failure lead to the overall instability of the column and significantly reduces the bearing capacity, which is 29.9% lower than that of the SQCC short column without hole. The bearing capacity of the short column is increased by 148.7%, 130.2%, and 134.3% respectively with the three reinforcement schemes of ASR-Angle steel reinforcement, PPR-Perforated Plate Reinforcement, and SPR-Surrounding Plate Reinforcement. The ASR scheme has the best reinforcement effect on the short column and the stress concentration decreases most obviously with this scheme, which is the most appropriate.

3.3 Bearing Mechanism of Joint Components

3.3.1 Experiment Scheme

3.3.1.1 Loading Scheme

The numerical tests of pure bending (four-point bending) are conducted with half the length of the components. The model size, the boundary conditions and the load conditions are shown in Fig. 3.15. Half of the casing length is l or θ, and the total length of the test specimen is 2.5 l or 2.5θ. According to the principle of symmetry, the boundary conditions of the specimen are determined:

(1) The right side is equivalent to sliding constraint, restricting the horizontal displacement and rotation of the casing. The sliding constraint is only applied to the casing cross-section, and right horizontal displacement of the right cross-section of the confined concrete component is limited by a baffle.
(2) Hinged constraints are applied on the left side to limit only the vertical displacement.
(3) The load point is in the middle position. No restriction is placed on other locations.

The numerical test scheme is designed, as shown in Table 3.4, to further study the mechanical behavior of casing joints and the effect of various parameters (concrete strength, casing length, casing thickness, installation clearance, etc.). The basic parameters of the components are listed in the table.

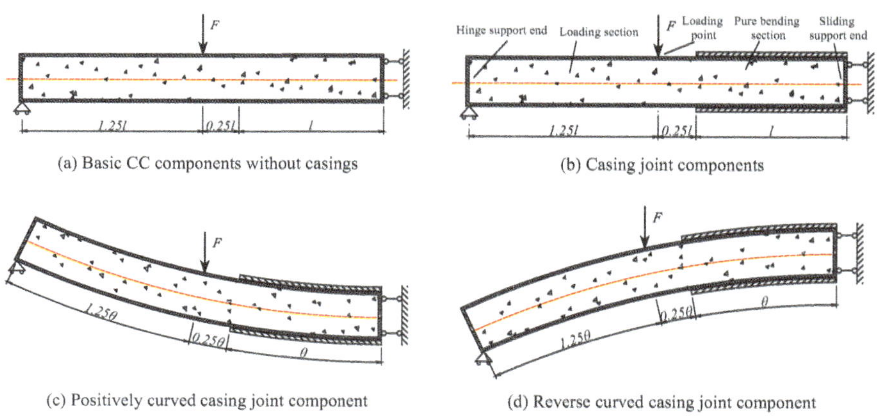

Fig. 3.15 The pure bending numerical test schemes for casing joints

Table 3.4 Finite element analysis results with experimental results

Sequence number	Test scheme	Test purpose	Critical curvature increment $\Delta\varphi_0$	M_{ue}/kN m	0.2 M_{ue}/kN m	$\Delta\varphi$ corresponded of 0.2 M_{ue}	0.6 M_{ue}/kN m	$\Delta\varphi$ corresponded of 0.6 M_{ue}	K_i/kN m^2	K_s/kN m^2
1	SQCC150*8-C40-0.45-straight (basic component)	Basic scheme	0	99.26	19.85	0.006	59.55	0.019	3240.98	3150.95
2	SQCC150*8-180*12-0.45-C40-straight (component with casing joint)		0.027	106.69	21.34	0.040	64.01	0.057	1604.36	2148.12
3	SQCC150*8-180*12-10°-C40-arc-R2.5 (positive bending)		0.025	108.04	21.61	0.038	64.82	0.052	1622.22	2364.11
4	SQCC150*8-180*12-10°-C40-arc-R2.5 (negative bending)		0.030	105.39	21.08	0.044	63.23	0.058	1548.71	2265.64
5	SQCC150*8-180*12-0.45-C30-straight	Influence mechanism of concrete strength	0.027	105.54	21.11	0.040	63.32	0.057	1587.07	2110.80
6	SQCC150*8-180*12-0.45-C50-straight		0.027	106.20	21.24	0.040	63.72	0.056	1609.09	2167.35
7	SQCC150*8-180*12-0.25-C40- straight	Influence mechanism of casing length	0.101	103.10	20.62	0.143	61.86	0.194	485.86	659.21
8	SQCC150*8-180*12-0.35-C40-straight		0.042	105.20	21.04	0.069	63.12	0.087	779.26	1394.30
9	SQCC150*8-180*12-0.55-C40- straight		0.017	102.35	20.47	0.027	61.41	0.040	2132.29	2706.48
10	SQCC150*8-182*12-0.45-C40-straight (δ = 4 mm)	Influence mechanism of casing space	0.035	104.14	20.83	0.049	62.48	0.068	1477.16	1884.32
11	SQCC 150*8-184*12-0.45-C40-straight (δ = 5 mm)		0.041	104.90	20.98	0.062	62.94	0.078	1049.00	1705.69
12	SQCC150*8-172*6-0.45-C40-straight (t = 3 mm)	Influence mechanism of casing thickness	0.028	104.73	20.95	0.054	62.84	0.074	819.82	1351.89
13	SQCC150*8-176*8-0.45-C40-straight (t = 3 mm)		0.027	104.39	20.88	0.040	62.63	0.056	1640.19	2144.83

(continued)

Table 3.4 (continued)

Sequence number	Test scheme	Test purpose	Critical curvature increment $\Delta\varphi_0$	M_{ue}/kN m	0.2 M_{ue}/kN m	$\Delta\varphi$ corresponded of 0.2 M_{ue}	0.6 M_{ue}/kN m	$\Delta\varphi$ corresponded of 0.6 M_{ue}	K_i/kN m^2	K_s/kN m^2
14	SQCC150*8-180*6-0.45-C40-straight–casing reinforce	Reinforce scheme of casing	0.0264	104.6	20.98	0.0368	62.94	0.0586	2017.31	1954.66
15	CCC159*8-0.45-C40-straight (without casing)	Comparison of circular components	0	82.9	16.58	0.00678	49.74	0.0212	2445.43	2346.23
16	CCC159*8-189*12-0.45-C40-straight		0.0234	91.3	18.26	0.0348	54.78	0.0492	1601.75	2123.26

Note Taking "SQCC 150*8-180*12-0.45-C40-arc-*R*2.5 (negative bending)" as an example, it is indicated that the edge length of the cross-section of SQCC is 150 mm, the wall thickness is 8 mm, the square steel pipe is filled with C40 concrete, and the square steel pipe with the cross-section edge length of 180 mm is used as the connecting member of the casing, the casing thickness is 12 mm, and the length of the casing is 0.45 m. R2.5 indicates that the radius of curvature of the arc component is 2.5 m, and other analogies

3.3.1.2 Model Settings

Based on the boundary conditions of the components identified in the previous section, the numerical component model is established as shown in Figs. 3.16 and 3.17. The vertical plate on the left is the constraining baffle, constraining three directions in U1(X), U2(Y), and U3(Z). The end of the component is in hard contact with the baffle. The friction force is 0; hinged constraint is applied to the other end of the component by limiting the vertical displacement on the horizontal midline of the cross-section.

3.3.2 Deformation and Failure Mechanism

The maximum effective stress curves of the steel and concrete core at different moments under the schemes 1–4 are shown in Fig. 3.18. The stress nephogram of the component under the ultimate load is shown in Figs. 3.19 and 3.20.

The analysis of the figures shows:

(1) In the initial state, there is a space between the casing and the confined concrete component. The rotation rigidity of the component is increased after a certain rotation. The rotation angle is defined as a critical angle of the casing joint.
(2) The deformation and stress values of the confined concrete components are significantly greater than that of the casings.
(3) For arch components with casing, the relationship between the component and the casing is the same as that of the straight casing under positive and negative bending conditions. The stress distribution is basically the same. The original curvature has little effect on the flexural bearing capacity and the stress state of the components.

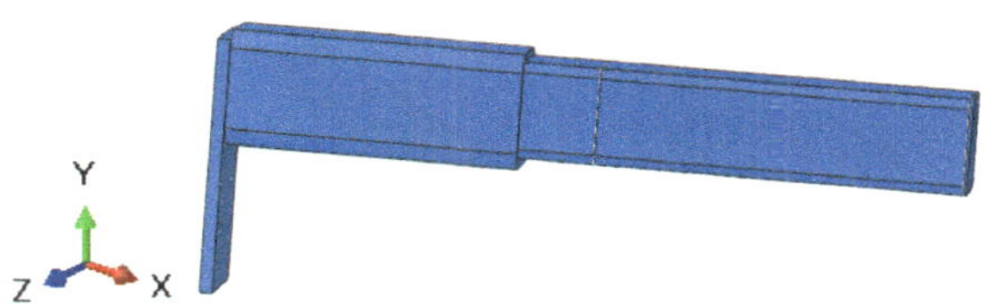

Fig. 3.16 Pure bending model

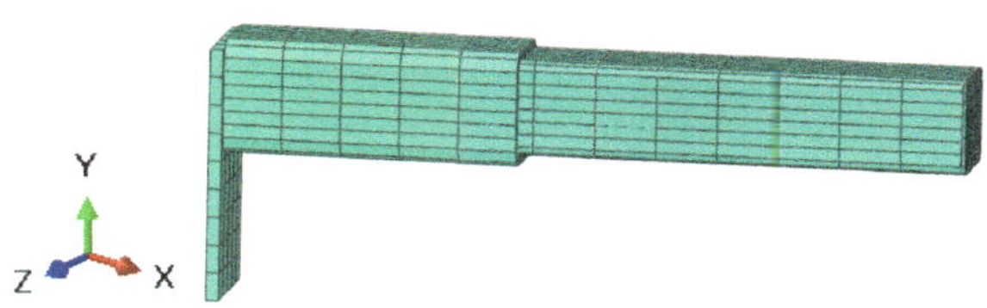

Fig. 3.17 Pure bending model mesh generation

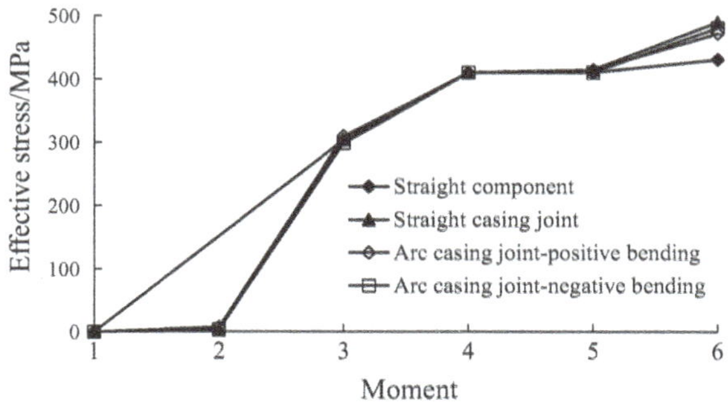

(a) Confining steel

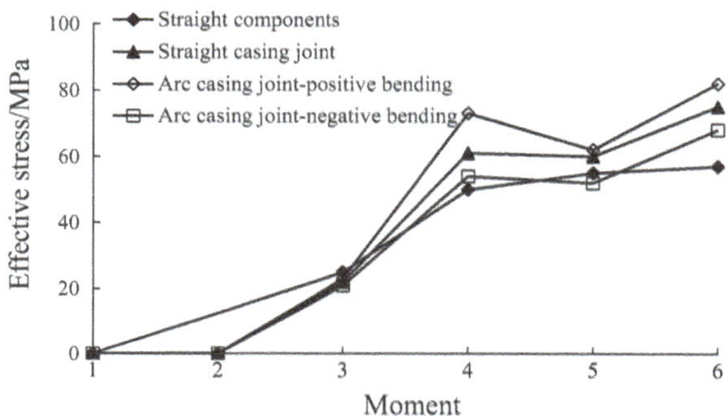

(b) Concrete core

Fig. 3.18 Maximum effective stresses of the components at different moments

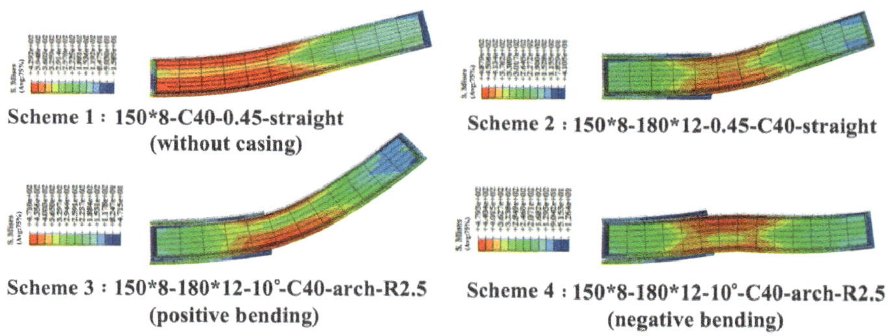

Fig. 3.19 Steel stress nephogram

(4) The concrete stress value in the scheme of the negative bending casing > the straight casing component > the positive bending casing component > the basic component without casing is the minimum. It shows that the negative bending is more unfavorable for the concrete core of the arc components.

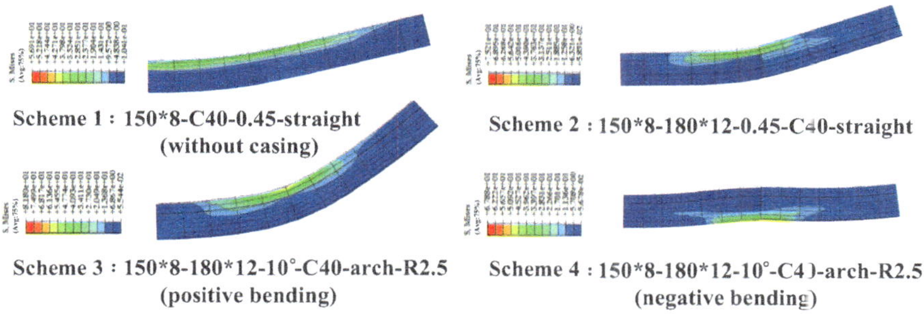

Fig. 3.20 Concrete stress nephogram

3.3.3 *Bearing Mechanism*

3.3.3.1 Parameter Definition

The relationship curve of the bending moment M of the cross-section of the casing joint and the curvature increment is defined as a $M - \Delta\phi$ curve which can directly reflect the flexural properties of the component. The value of $\Delta\phi$ is positive as the curvature is increased and negative as the curvature is decreased.

Figure 3.21 is the $M - \Delta\phi$ curve of the basic test scheme (scheme 1–4), and Figs. 3.22 and 3.23 are the typical $M - \Delta\phi$ curves of the basic confined concrete components and the joint components. The definition is:

(1) Point O indicates the loading starting point. Point A is the critical point of the hinged-rigid joints. Point B is the corresponding position of 0.2 M_{ue}. Point C is the corresponding position of 0.6 M_{ue}. Point D is the end of the elastic stage. Point E is the middle of the curve inflection point (middle of elastic stage and plastic stage). Point F is the corresponding position of M_{ue}. Point G represents

Fig. 3.21 Numerical test curve

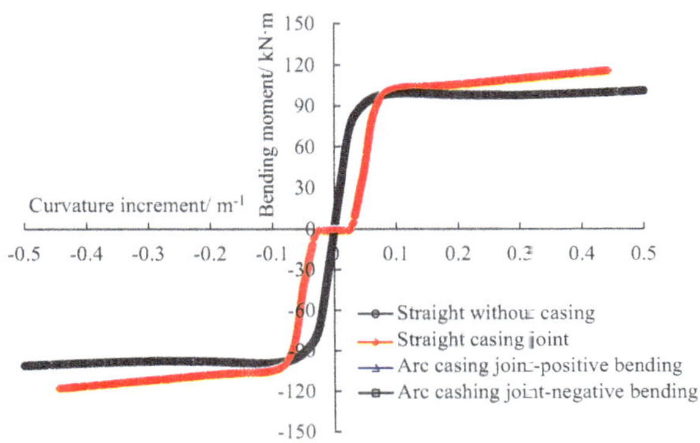

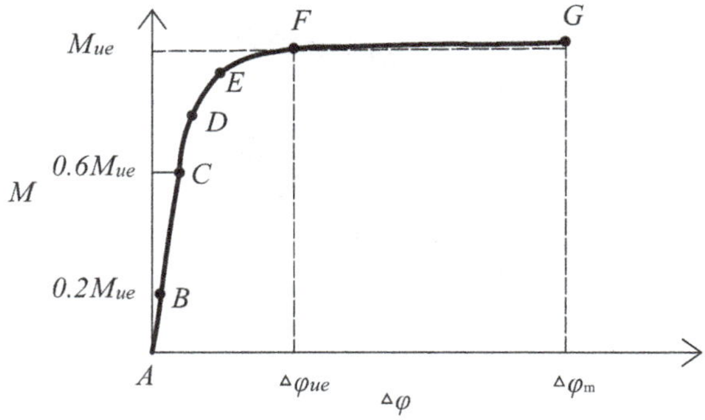

Fig. 3.22 Basic component curve

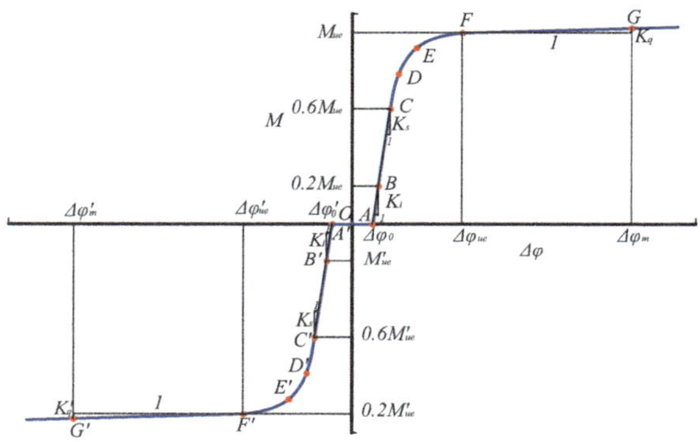

Fig. 3.23 Joint component curve

the end of the curve ($\Delta\phi = 0.2$ is set as the component without the casing, and $\Delta\phi = 0.4$ is set as the component with the casing).

(2) The effective length l_0 of the casing joint: As shown in Fig. 3.15, the effective length l_0 of the casing joint in this paper is 2.5 l or 2.5θ, where l or θ is half of the actual length of the casing joint with the unit as m.

(3) Ultimate bending moments M_{ue}: The bending moment value corresponding to the designed curvature increment.

(4) Curvature increment $\Delta\phi$: the curvature change of the pure bending section during the test and the unit as m^{-1}.

(5) Critical curvature increment $\Delta\phi_0$: the curvature increment corresponding to point A in Fig. 3.23, that is the curvature increment when the casing joint changes from the hinged state to the rigidly connected state, and the unit is m^{-1}.

(6) Critical angle ω_0: the rotation angle of the casing joint which changes from the hinged state to the rigidly connected state with the unit as rad.

(7) Equivalent rigidity of the casing joint: It reflects the flexural properties of the components within the effective length of the casing joint. The corresponding component range is l_0 and the rigidity is divided into K_i and K_s.

(8) Equivalent rigidity K_i in the initial stage: $K_i = \frac{0.2M_{ue}}{\Delta\phi_{0.2}}$, K_i is the secant rigidity at the position of $M = 0.2\ M_{ue}$ with the unit as N m^2.
(9) Equivalent rigidity K_s in bearing stage: $K_s = \frac{0.6M_{ue}}{\Delta\phi_{0.6}}$, K_s is the secant rigidity at the position of $M = 0.6\ M_{ue}$ with the unit as N m^2.

The ultimate bending moment and equivalent rigidity of each specimen are shown in Table 3.4.

3.3.3.2 The Bearing Mechanism Analysis

1. The Basic Confined Concrete Component

The $M - \Delta\phi$ curve form of the basic confined concrete component without casing joints can be divided into three stages:

(1) Elastic stage (OD): In the initial loading stage, the specimen is in the elastic deformation stage, and the growth rate of curvature is much lower than that of the external load.
(2) Elastic-plastic stage (DF): Afterwards, the steel enters the yielding stage, the growth rate of the component deformation is faster than that of the external load.
(3) Plastic stage (FG): As the loading continues, the bearing capacity of the component continues to increase slowly. This is mainly due to the composite effect between the steel pipe and the concrete. The component has a better ductility.

2. Casing Joint Components

The bearing properties of the casing joint mainly include:

(1) Due to the critical angle of the casing joint, the component $M - \Delta\phi$ curve has an additional OA stage in the initial stage of loading. The component does not transfer the bending moment, and the casing joint is equivalent to a hinge joint at this stage.
(2) The curve shape of the casing joint after the OA stage is basically the same as that of the basic component. The OA stage is followed by the elastic stage (AD), the elastic-plastic stage (DF) and the plastic stage (FG). The curvature increment corresponding to point A is the critical curvature increment $\Delta\phi_0$. When the curvature increment exceeds this value, the component enters the rigid state and begins to transfer the bending moment.
(3) For the straight casing joint, there is no difference between the positive and the negative bending, so the $M - \Delta\phi$ curves in the negative bending state and in the positive bending state are antisymmetric. They also undergo 4 stages.
(4) The ultimate bending moment of the component with casing joint is 7.5% higher than that of the basic confined concrete component. Correspondingly, the rigidity in the initial stage and using stage is reduced by 50.5 and 31.8%.
(5) The rigidity K_i of the basic confined concrete component in the initial stage is slightly larger than the rigidity K_s in the using stage by 2.86%; but the rigidity K_i

of the component with casing joint in the initial stage is about 20–30% smaller than the rigidity K_s in the using stage. The main reason is that the load of the basic component is entirely born by the confined concrete component. The rigidity difference is not significant between the initial stage and the bearing stage; due to the small contact area between the casing and the confined concrete component in the initial stage of loading, stress concentration and local deformation occur on the casing; and as a result, the overall rotation angle of the component is enlarged and the initial rigidity is smaller. In the past stage of loading, the interaction between the casing and the confined concrete component is sufficient, and the contribution rate of the local deformation to the rotation angle is reduced. As a result, the rigidity K_s is significantly larger in the bearing stage than in the initial stage and it is closer to that of the basic component.

3. Arc Joint Components

Since the arc joint component has a certain original curvature, there is a difference between positive and negative bending, as shown in Fig. 3.23. The following conclusions can be drawn from Fig. 3.21 and schemes 2–4 in Table 3.4:

(1) There is a difference in the mechanical parameters between the arc casing joint and the straight casing joint. In addition, there are differences in the mechanical parameters of the arc casing joint under the conditions of the positive and the negative bending.

(2) Critical curvature increment $\Delta\phi_0$: the negative bending arc casing joint component > the straight casing joint component > the positive bending arc casing joint component.

Ultimate bending moment M_{ue}: the positive bending arc casing joint component > the straight casing joint component > the negative bending arc casing joint component.

Rigidity K_i in the initial stage: the straight casing joint component > the negative bending arc casing joint component > the positive bending arc casing joint component.

Rigidity K_s in the bearing stage: the positive banding arc casing joint component > the negative bending arc casing joint component > the straight casing joint component. In a comprehensive comparison, the arc casing joint has better flexural properties under the positive bending conditions.

(3) Although there are differences in the results of the above three schemes, the differences are not significant. Except that the bending rigidity difference rate is 10% in the bearing stages, all the other difference rates are within 5%. Therefore, the conclusion of the straight component can apply to the arc components without causing significant errors.

3.3.3.3 The Influence Mechanism

1. The Influence Mechanism of Concrete Strength Grade

Figure 3.24 shows the curve $M - \Delta\phi$ of the component and the bending rigidity curves for different concrete strength grades.

The curve analysis shows:

(1) The concrete strength grade has almost no effect on the critical curvature increment and the $M - \Delta\phi$ curve form, and has little effect on the ultimate bending moment.

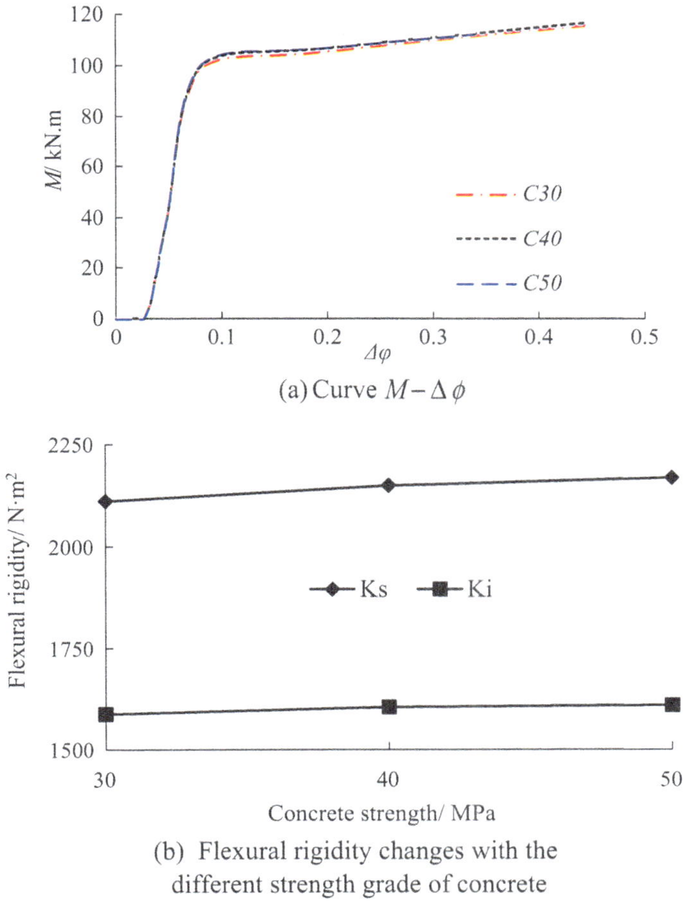

Fig. 3.24 Flexural properties in the different concrete strength grades

(2) The bending rigidity shows an approximately linear increase with the increase of the concrete strength grade. The effect is more obvious on the flexural rigidity K_s in the bearing stage.

2. Influence Mechanism of the Casing Length

The influence mechanism of the casing length l on the flexural properties of the component is shown in Fig. 3.25.

The curve analysis shows:

(1) The length of the casing has a significant influence on the flexural properties and the curve $M - \Delta\phi$. The longer the casing length is, the larger the slope of the curve is, and the curve gradually approaches the component without casing.
(2) The length of the casing has little influence on the ultimate bending moment of the component which increases first and then decreases as the length of the casing increases.
(3) The critical angle shows a significant decrease with the increase of the casing length. The longer the casing is, the smaller the range of the component in the hinged state.
(4) In the initial stage, the flexural rigidity and the casing length have an approximately exponential positive correlation. In the bearing stage, the flexural rigidity has an approximately linear and positive relationship with the casing length.

3. Influence Mechanism of the Casing Wall Thickness

The effect curves of the casing wall thickness t on the flexural properties of the component is shown in Fig. 3.26.

The analysis on the influence mechanism of the casing wall thickness shows:

(1) When the casing wall thickness is 8 mm or more, the wall thickness variation has little influence on component curve $M - \Delta\phi$, critical angle and bending rigidity. When the casing wall thickness is less than 8 mm, it has a significant influence on the flexural properties of the component.
(2) The wall thickness of the casing has no obvious effect on the ultimate bending moment of the component with an slight growth trend. It has an obvious effect on the post bearing capacity. The thicker the wall thickness is, the higher the post bearing capacity is.
(3) The casing wall thickness has almost no effect on the critical angle of the component.
(4) The wall thickness of the casing has a significant influence on the flexural rigidity of the component. The flexural rigidity with a wall thickness of 8 mm is 1.62 times that with a wall thickness of 6 mm. The casing corresponding flexural rigidity is basically equal no matter whether the wall thickness is 12 or 8 mm.
(5) Comparing with the original scheme 12, the casing reinforcement scheme 14 (welding around both the ends of the casing of $t_0 = 6$ mm stiffener plate with

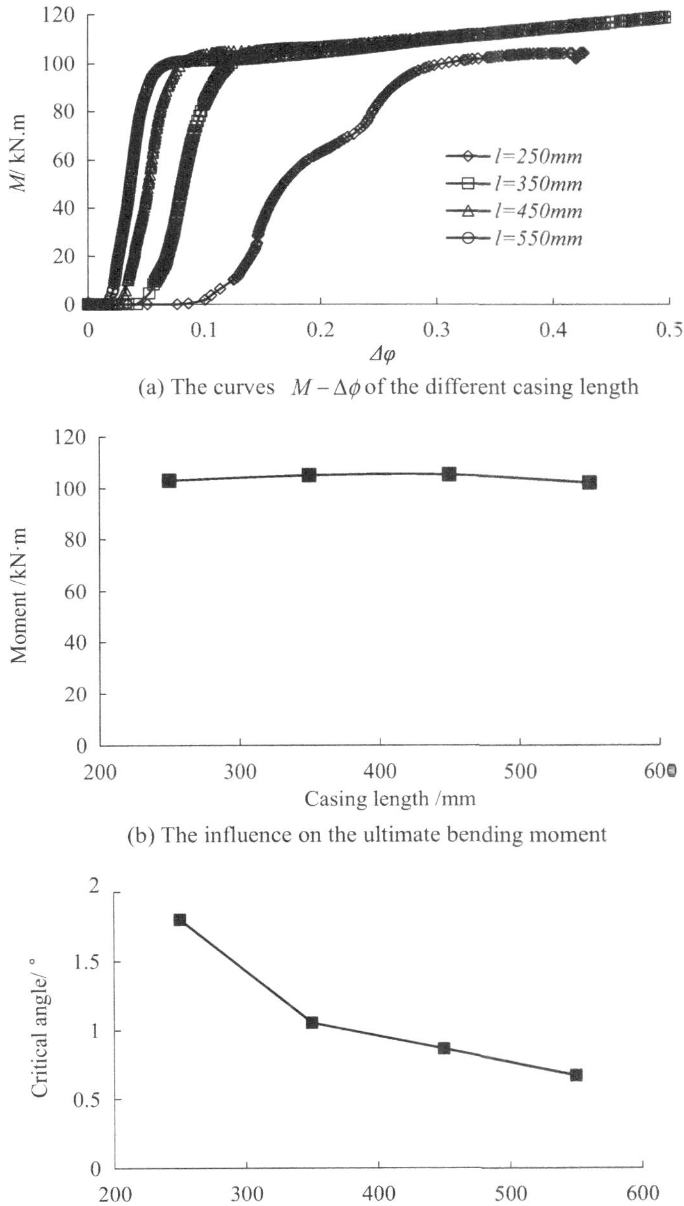

(a) The curves $M-\Delta\phi$ of the different casing length

(b) The influence on the ultimate bending moment

(c) The influence on the critical angle

Fig. 3.25 The influence of the different casing length on the bending performance of the components

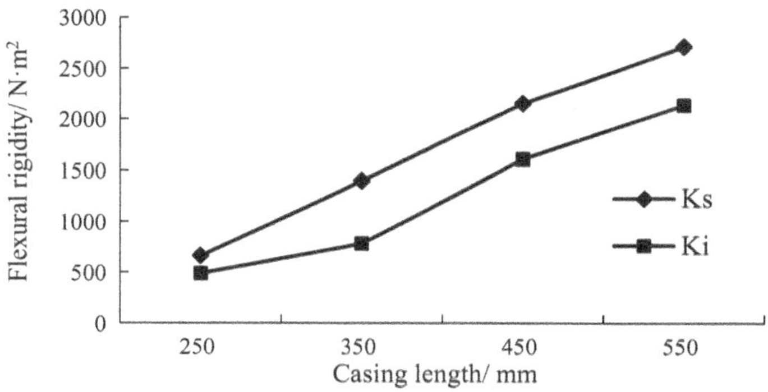

(d) The influence on the flexural rigidity

Fig. 3.25 (continued)

a thickness of 8 mm and a width of 50 mm) basically has almost no effect on the critical angle and ultimate bending moment of the joint. It significantly influences the flexural rigidity which increases by 146% to 2019.3 kN m^2 in the initial stage and by 44.6% in the bearing stage. The flexural rigidity in the bearing stage exceeded 1604.4 kN m^2 in the scheme 2. The casing reinforcement scheme in this paper provides an effective way to improve the flexural rigidity of the casing joint.

4. Influence Mechanism of the Casing Space

The effect curves of the space between casings on the flexural properties of the component is shown in Fig. 3.27.

The analysis on the influence mechanism of the casing space shows:

(1) The casing space has an obvious influence on the curve $M-\Delta\phi$ and the length of stage without bending moment increases with the increase of the casing space.
(2) With the increase of the casing space, the ultimate bending moment of the components slightly decreases.
(3) With the increase of the casing space, the critical angle of the components increases linearly, and the influence is significant.
(4) With the increase of the casing space, the flexural rigidity of the components decreases linearly, and the influence is significant.

3.3.4 Summary

The numerical comparison tests on the basic confined concrete components with and without casing joint clarify the deformation and failure mechanism and bearing mechanism of the components with casing join. The influence of the various factors

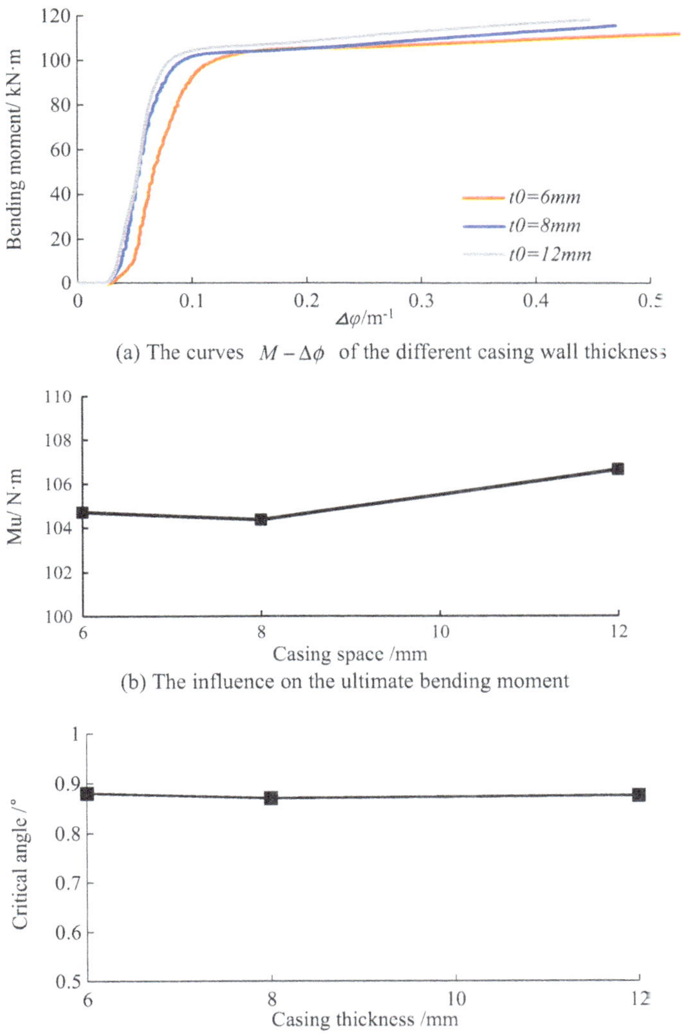

(a) The curves $M-\Delta\phi$ of the different casing wall thickness

(b) The influence on the ultimate bending moment

(c) The influence on the critical angle

Fig. 3.26 The influence of the wall thickness on the flexural properties of components

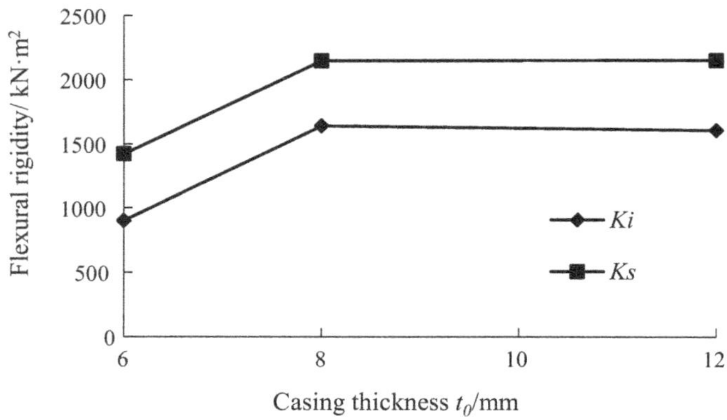

(d) The influence on the flexural rigidity

Fig. 3.26 (continued)

is obtained on the mechanical properties of components. The factors include the concrete grade, the casing length, the casing thickness, the casing space and etc.

3.4 Chapter Summary

1. Axial Bearing Mechanism Test of the Basic Component

The SQCC short column shows a plastic instability failure mode and it avoids the bending instability as the traditional U-steel short column shows. Its typical N–ε curve shows four stages from rapid rise to gentle rise without fall stage. This indicates the SQCC short column has a good ductility and post bearing capacity.

2. Component Reinforcement with Grouting Holes

Compared with the SQCC short column without hole, the strength of the short column with grouting hole is decreased by 29.9%. The bearing capacity of the short column is increased by 148.7%, 130.2%, and 134.3% respectively with the three reinforcement schemes of ASR-Angle steel reinforcement, PPR-Perforated Plate Reinforcement, and SPR-Surrounding Plate Reinforcement. The ASR scheme has the best reinforcement effect on the short column and the stress concentration decreases most obviously with this scheme, which is the most appropriate.

3. Bearing Mechanism of Joint Components

The pure bending numerical tests on the casing joint components clarify the deformation and failure mechanism and the bearing mechanism of the confined concrete casing joint. The influence of different factors is analyzed on the mechanical properties.

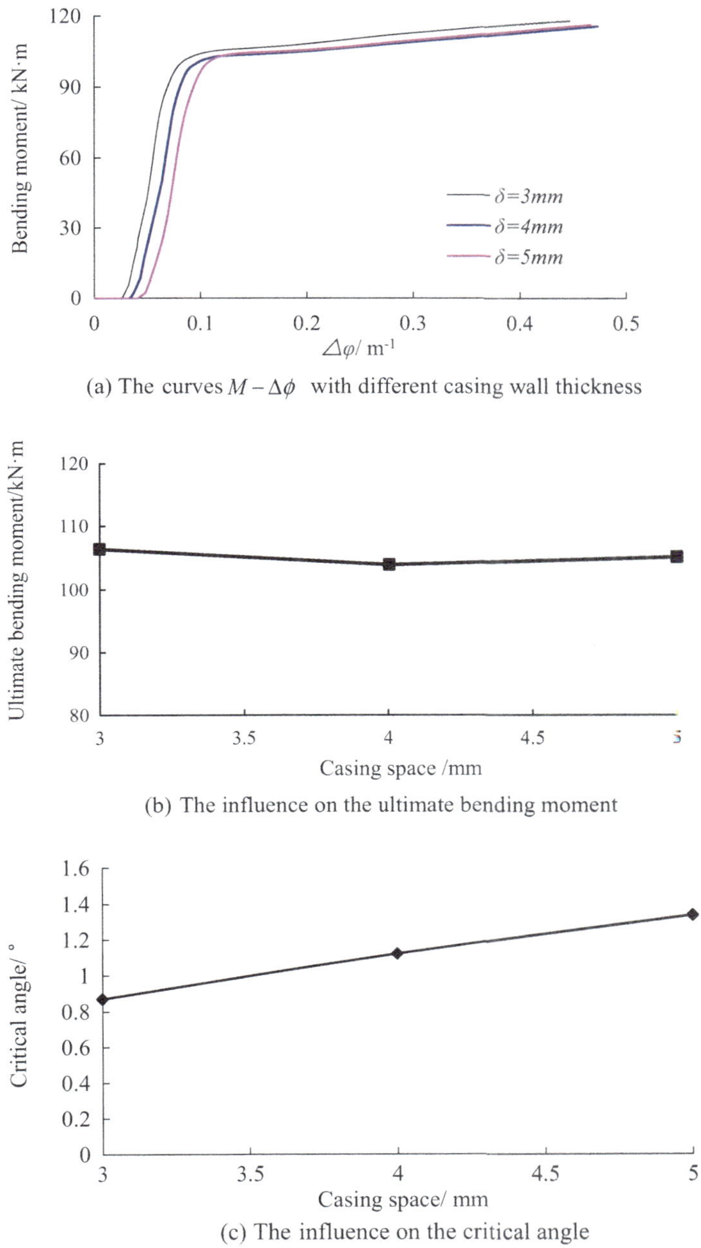

Fig. 3.27 The influence of the casing gap on the flexural properties of the component

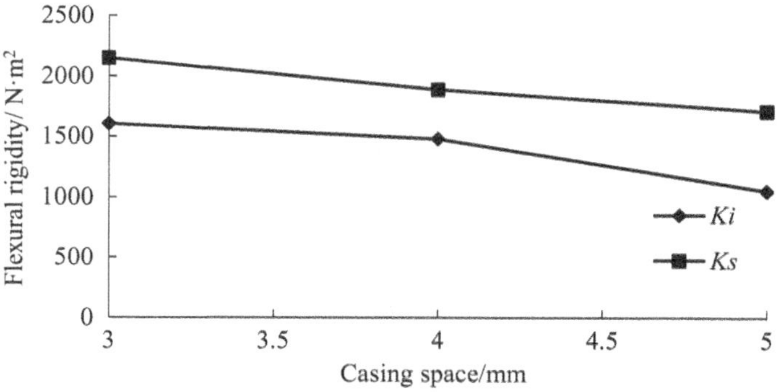

(d) The influence on the flexural rigidity

Fig. 3.27 (continued)

Reference

Han LH (2019) Concrete fill steel tubular structures-theory and practice. Science Press, Beijing

Chapter 4
Calculation Theory of the New High Strength Arch

As mentioned above in this book, confined concrete arch is a new type of high strength support form, but there is no reasonable calculation theory for the confined concrete arches in roadways. This chapter first establishes the calculation and analysis model for the casing joint of the arch, and analyzes the action mode and mechanical properties of the casing joint. The calculation formulas of internal force are derived for arbitrary section numbers and unequal rigidity of straight-wall semi-circular and circular arches. Combined with the bending strength criterion of confined concrete components, the arch internal force state and bearing capacity are calculated and analyzed.

4.1 Symbol Description

A	Cross-section area of the I-steel
A_a	Axis force calculation area of I-steel
A_c	Cross-section area of concrete
A_f	Flange area of I-steel
A_s	Cross-section area of steel pipe
A_{sc}	Cross-section area of confined concrete, $A_{sc} = A_s + A_c$
A_w	Web area of I-steel
B	The outer length of the square steel pipe cross section or the length (width) of the short edge of the rectangular steel pipe cross section
b	Width of the I-steel cross-section
B_0	Casing length

(continued)

Q. Wang et al., *High Strength Support for Soft Surrounding Rock in Deep Underground Engineering*, https://doi.org/10.1007/978-981-15-3844-5_4

(continued)

D	The outer diameter of the circular steel pipe cross section or the length (height) of the long side of the rectangular steel pipe cross section
e	Axial load eccentricity
e/r	Load eccentricity
E_c	Elastic modulus of concrete
E_s	Elastic modulus of steel
E_{sc}	Axial elastic modulus of confined concrete
EI	Flexural rigidity
EI'	Equivalent rigidity of joints
EI_T	Flexural rigidity of casing cross section
f_{ck}	Standard values of concrete axial compressive strength
f'_c	Compressive strength of concrete cylinder
f_{cu}	Compressive strength of concrete cube
f_{sc}	Compressive strength of the steel pipe filled with concrete
f_{scy}	Strength index of confined concrete under axial compression
f_u	Ultimate tensile strength of steels
f_y	Yield strength of steels
H	Straight leg height of straight-wall semi-circle arch
h	Height of the I-steel cross-section
I_c	Bending moment of inertia of core concrete cross-section
I_s	Bending moment of inertia of steel pipe
I_{sc}	Bending moment of inertia of confined concrete
K_i	Bending moment at initial stage
K_s	Bending moment at service stage
l	Casing centerline arc length
l_0	Effective length or calculation length of casing joint or clamp
M	Bending moment
M_u	Ultimate bending moment
M_{ue}	Experimental value of ultimate bending moment
N	Axial pressure (compressive pressure is positive in this book)
N_u	Ultimate axial force
N_{ue}	Experimental values of ultimate axial force
q	Pressure acting on arch
q_{1cr}	Ultimate bearing capacity of arch
r	Fillet radius of cold formed steel pipe corner
R	Centerline radius of arch
R'	Outer radius of the casing
t	Wall thickness of steel pipe of confined concrete

(continued)

(continued)

t_0	Wall thickness of casing
t_f	Flange plate thickness of I-steel
t_w	Web plate thickness of I-steel
W_{scm}	Bending modulus of component cross-section
α	Joint orientation angle or cross-section steel content of confined concrete ($\alpha = A_s/A_c$)
β_m	Equivalent moment coefficient
γ	Angle and radian between the arbitrary cross-section and the end face of the arch casing joint
γ_m	Calculation coefficient of bending rigidity bearing capacity ($\gamma_m = M_u/W_{scm}f_{scy}$)
2δ	Total gap between confined concrete component and casing The flexure operation to increase the arch curvature is positive curve, and the contrary is negative curve
ε	Strain
θ	Angle between the two ends of the arch casing joint, it is used to describe the dimension and the radian of casing in the longitudinal direction
κ	High-diameter ratio
λ	Slenderness ratio of confined concrete component or lateral pressure coefficient
μ	Equivalent joint rigidity ratio
ξ	Confinable effect coefficient ($\xi = A_s f_y/A_c f_{ck}$)
σ	Material stress
φ	Angle and radian between the arbitrary cross-section of the casing joint and the central cross-section
ϕ	Curvature
$\Delta\phi$	Increment of curvature
$\Delta\phi_0$	Critical curvature increment
ω	Joint rotation angel
ω_0	Joint critical rotation angle

4.2 Arch Calculation Model

The cross-sectional shape of most roadways is straight-wall semi-circular and circular, and the cross-sectional shape of most confined concrete arches is square and circular, as shown in Fig. 4.1.

Mechanical analysis models are established as straight-wall semi-circular and circular arches by simplifying the main structure, load and constraints of the arches.

Structural simplification: The main structure of the arch is simplified as a curve along the axis of the arch. The flexural rigidity of the axis is EI. The length of the joint is taken as l_0. The equivalent rigidity of the joint is taken as EI'.

Fig. 4.1 Cross-sectional shape of arches

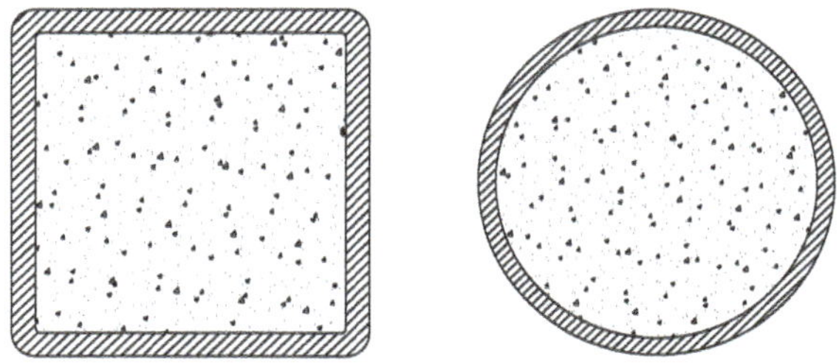

(a) Square cross-section (b) Circular cross-section

Constraint simplification: The arch bottom (point A) can be simplified as a hinge support due to the action of feet-lock bolt. According to symmetry, the vault (point C) of the straight-wall semi-circular arch can only move up and down, so the vault is simplified as a directional support. The vault and the arch bottom of the circular arch are also simplified in the same way. A casing joint is located at point B and its simplification is determined by its mechanical properties.

Load simplification: The load is simplified as a vertical load q_1 and a horizontal load q_2.

The arch calculation model studied in this book is shown in Fig. 4.2.

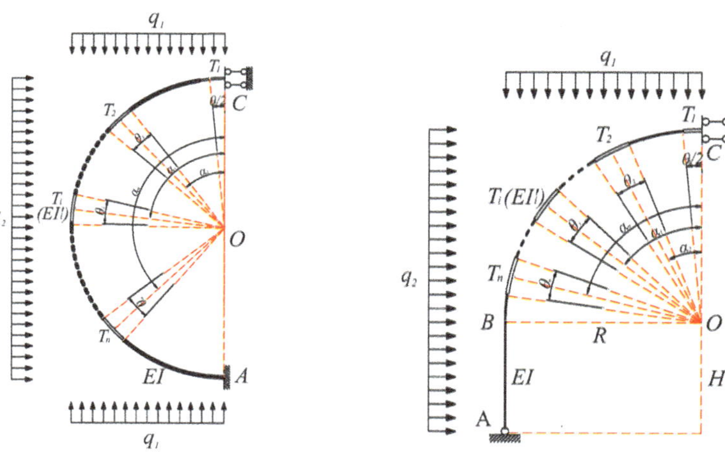

(a) Circular arches with arbitrary sections (b) Straight-wall semi-circular arches with arbitrary sections

Fig. 4.2 Mechanical analysis model of arch

4.3 Arch Joint Calculation Analysis

4.3.1 Arch Joint Action Mode

1. Analysis Object

Three cross-sections of straight-wall semi-circular arches are taken as examples. The casing joints are analyzed. Two joints are symmetrically arranged on the left and right sides of the arch. The position of the joint is represented by the angle α between the center cross-section of the joint and the central axis of the arch, as shown in Fig. 4.3.

The physical meaning of each parameter:

- R Radius of the arch centerline, the inner radius of the casing, mm;
- R' Outer radius of the casing, mm;
- H Straight leg height of the arch, mm;
- α Angle between the center cross-section of the joint and the central axis of the arch, it is used to describe the position and the radian of the joint.
- θ Angle between the two ends of the casing joint, it is used to describe the dimension and the radian of casing in the longitudinal direction;
- φ Angle and radian between the arbitrary cross-section of the casing joint and the central cross-section;
- γ Angle and radian between the arbitrary cross-section and the end face of the arch casing joint;
- B_0 Outer casing dimensions, edge length, mm;
- B Outer dimensions of the arch components, edge length, mm;
- t_0 Casing wall thickness, mm;
- 2δ Total space between confined concrete components and casing, mm;
- l Arc length of casing center line, mm;

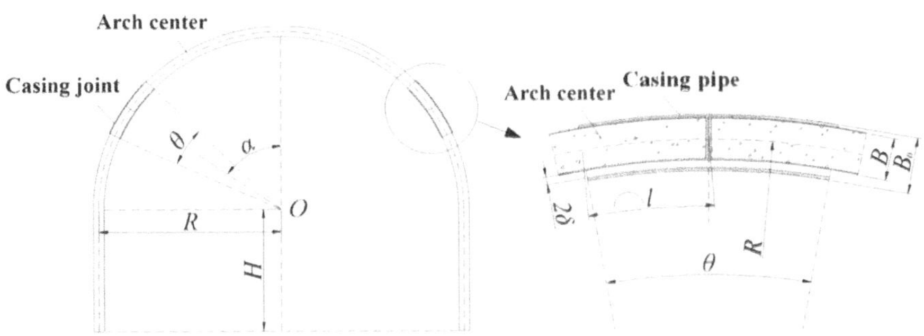

Fig. 4.3 Casing joint and its dimensions

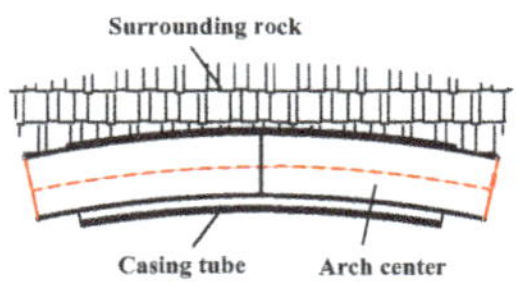

(a) Hinge joint mode

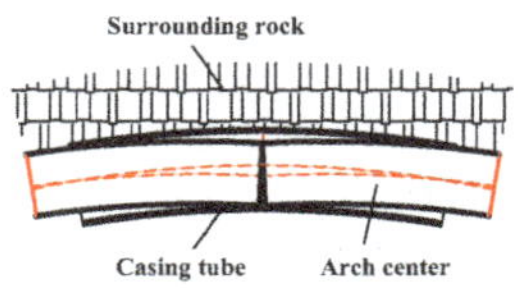

(b) Rigid joint mode (concave, negative bending)

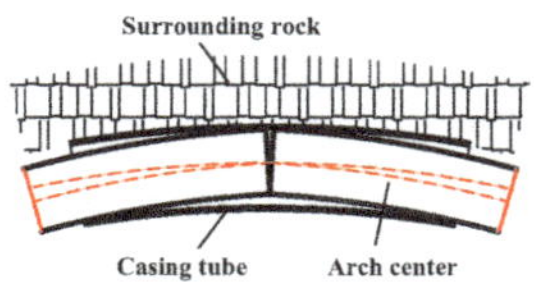

(c) Rigid joint mode (convex, positive bending)

Fig. 4.4 Schematic diagram of the role of the casing joint

2. Analysis on the Model and the Casing Action Mode

At the initial stage of the arch installation, the outer edge of the arch is close to the casing due to the restriction of the surrounding rock. The main working mode of the casing is shown in Fig. 4.4a. The arches are approximately hinged, and the axial force and partial shear forces can be transferred between arches due to the restriction of the casing, which is called the hinged joint mode.

With the increase of the surrounding rock pressure, the arches are deformed and the convex or concave phenomenon occurs at the joints. The conditions that most likely occur on the casing joints are shown in Fig. 4.4b, c. The final working mode of the joint is changed from the hinge joint mode to the rigid joint mode. Both the axial force and the shear force can be transmitted between the components, and the bending moment of the components can be transmitted as well. The flexural rigidity is represented by EI'.

4.3.2 *Casing Force Analysis*

4.3.2.1 Mechanical Model

In the final working mode of rigid joint, the casing mechanical model is obtained through simplification, as shown in Fig. 4.5. The simplified description is as follows:

(1) The acting force on the casing by surrounding rocks is simplified as the uniform load q along the casing outer contour.
(2) The acting force on the contact position between the arch and the casing is simplified as a concentrated load. In the center of the casing, the shear forces F_{A2} and F_{B2} are transmitted on the casing by the arch. The position and the force direction of F_{A2} and F_{B2} acting on the casing are almost the same. Therefore, it is simplified as a concentrated load and denoted by F_2, $F_2 = F_{A2} + F_{B2}$.
(3) F_{A1}, F_{B1}, F_{A2}, and F_{B2} are the shear forces transmitted by the arches to the casing.

The simplified model of the final mechanical analysis is shown in Fig. 4.6.

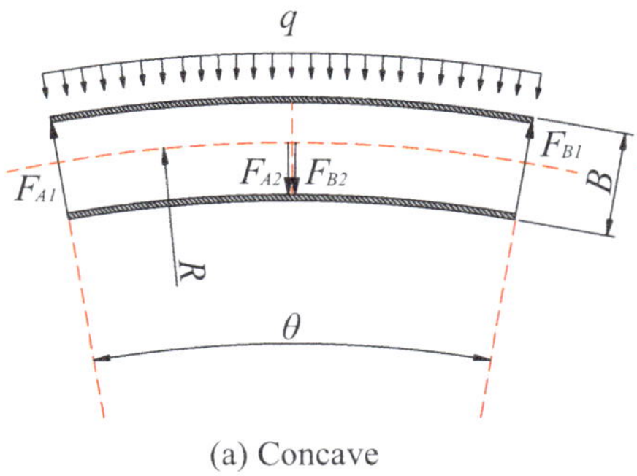

Fig. 4.5 Casing mechanical model in rigid mode

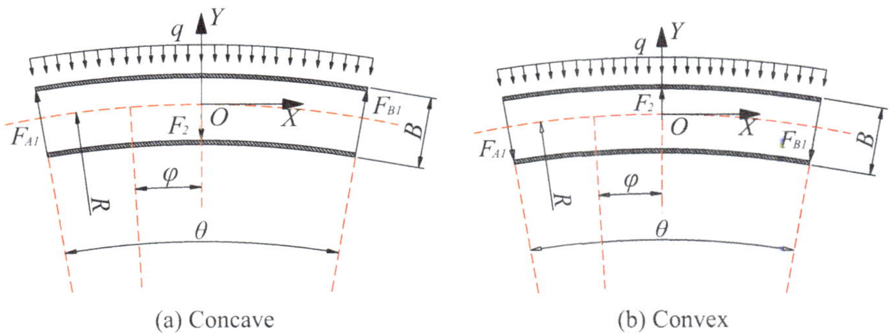

Fig. 4.6 Casing mechanical model in rigid joint mode (final model)

4.3.2.2 External Force Analysis

According to the calculation and analysis in Fig. 4.6, first, the symmetry of the structure shows:

$$F_{A1} = F_{B1} = F_1 \tag{4.1}$$

The equilibrium formula of the force in the concave state:

$$F_2 = F_{A1} \cdot \cos\frac{\theta}{2} + F_{B1} \cdot \cos\frac{\theta}{2} - 2\int_0^{\theta/2} q \cdot R' \cos\varphi d\varphi \tag{4.2}$$

$$F_2 = 2F_1 \cdot \cos\frac{\theta}{2} - 2qR' \sin\frac{\theta}{2} \tag{4.3}$$

The equilibrium formula of the force in the convex state:

$$F_2 = F_{A1} \cdot \cos\frac{\theta}{2} + F_{b1} \cdot \cos\frac{\theta}{2} + 2R' \int_0^{\theta/2} q \cdot \cos\varphi d\varphi \tag{4.4}$$

$$F_2 = 2F_1 \cdot \cos\frac{\theta}{2} + 2qR' \sin\frac{\theta}{2} \tag{4.5}$$

In the specific calculation, the magnitude F_1 of the shearing force F_{A1} and F_{B1} transmitted by the arches to the casing can be calculated by the overall stress of the arches. Based on this, the unknown force F_2 of the casing can be calculated by formulas (4.3) and (4.5).

4.3.2.3 Internal Force Analysis

1. Axial Force

The axial force in each location of the casing is 0.

2. Bending Moment of the Angle φ Between Arbitrary Cross-section of the Casing Joint and the Center Cross-section

In concave state:

$$M_\varphi = F_1 \cdot \sin\left(\frac{\theta}{2} - \varphi\right) - \int_0^{\frac{\theta}{2}-\varphi} qRR' \sin\gamma d\gamma \tag{4.6}$$

$$M_\varphi = F_1 \cdot \sin\left(\frac{\theta}{2} - \varphi\right) - qRR'\left(1 - \cos\left(\frac{\theta}{2} - \varphi\right)\right) \tag{4.7}$$

In convex state:

$$M_\varphi = -F_1 \cdot \sin\left(\frac{\theta}{2} - \varphi\right) - \int_0^{\frac{\theta}{2}-\varphi} qRR' \sin\gamma d\gamma \tag{4.8}$$

$$M_\varphi = -F_1 \cdot \sin\left(\frac{\theta}{2} - \varphi\right) - qRR'\left(1 - \cos\left(\frac{\theta}{2} - \varphi\right)\right) \tag{4.9}$$

where, R' is the outer radius of the casing, $R' = R + B_0/2 + t_0$.

After the internal force of the casing is obtained, the strength can be checked according to the size and the mechanical parameters of the casing. It should be mentioned that the specific behavioral mode is more complex since the interaction between the casing and the confined concrete component is a dynamic process. Therefore, a more accurate strength can be checked by experiments or numerical calculations.

4.3.2.4 Design and Check of Rigidity of Casing Cross-Section

The cross-section of the casing is square with an edge length of B_0 and a wall thickness of t_0, and its flexural rigidity EI_T is:

$$EI_T = \frac{E_s}{12}\left(B_0^4 - (B_0 - 2t_0)^4\right) \quad (4.10)$$

The flexural rigidity EI of the confined concrete component can be calculated by the following formula.

$$EI = E_s \cdot I_s + 0.2E_c \cdot I_c \quad (4.11)$$

In the above two formulas,

E_s is steel elastic modulus, generally taking 2.06×10^5 MPa; I_s is inertia moment of the cross-section of the confining steel pipe; E_c is concrete elastic modulus, $E_c = 21000\sqrt{f_c'/19.6}$ unit in MPa, f_c' is the compressive strength of the concrete cylinder, and $f_c' = 0.79 f_{cu,k}$, $f_{cu,k}$ is the standard value of the compressive strength of the concrete cube; I_c is cross-sectional inertia moment of the concrete.

In the design of the confined concrete arch casing, the flexural rigidity of casing cross-section can be checked preliminarily by formula (4.12):

$$EI_T \geq EI \quad (4.12)$$

The condition ensures that the flexural properties of the casing is not less than that of the confined concrete component under the bending moment.

Example 1 The edge length of an SQCC component is 150 mm, the wall thickness is 8 mm, and the concrete strength grade is C40; the outer diameter of the casing is 180 mm and the wall thickness is 12 mm.

The flexural rigidity of the confine concrete components is:

$$\begin{aligned} EI &= 2.06 \times 10^{11} \cdot ((0.15^4 - 0.134^4)/12) + 0.2 \times 0.21\sqrt{0.79 \times 40/19.6} \\ &\quad \times 10^{11} \times (0.134^4/12) \\ &= 3300\,\text{kN m}^2 \end{aligned}$$

The flexural rigidity of the casing cross-section:

$$\begin{aligned} EI_T &= 2.06 \times 10^{11} \cdot \left((0.18^4 - 0.156^4)/12\right) \\ &= 7854\,\text{kN m}^2 \end{aligned}$$

$EI_T > EI$, therefore, the flexural rigidity of the casing cross-section meets the design requirements.

It should be noted, with the flexural rigidity calculation formula of the casing joint, only the theoretical value of the flexural rigidity of the casing joint is obtained. In practical engineering, the interaction between the casing joint and the confined concrete component is not ideal, and stress concentration and local deformation occur. The result in the actual flexural rigidity of the casing is less than the value, so it is necessary to use other methods for further calculation.

4.4 Arch Internal Force Calculation

4.4.1 Internal Force Calculation of Circular Arch

4.4.1.1 Calculation Methods and Mechanical Model

The loads are equivalent to the horizontal and vertical uniform loads, respectively. The arches are equivalent to a curve along the axis line with the flexural rigidity as EI. The casing joint is equivalent to the arc with the flexural rigidity as EI'. After being simplified, the arch mechanical model is shown in Fig. 4.7. The common arches at

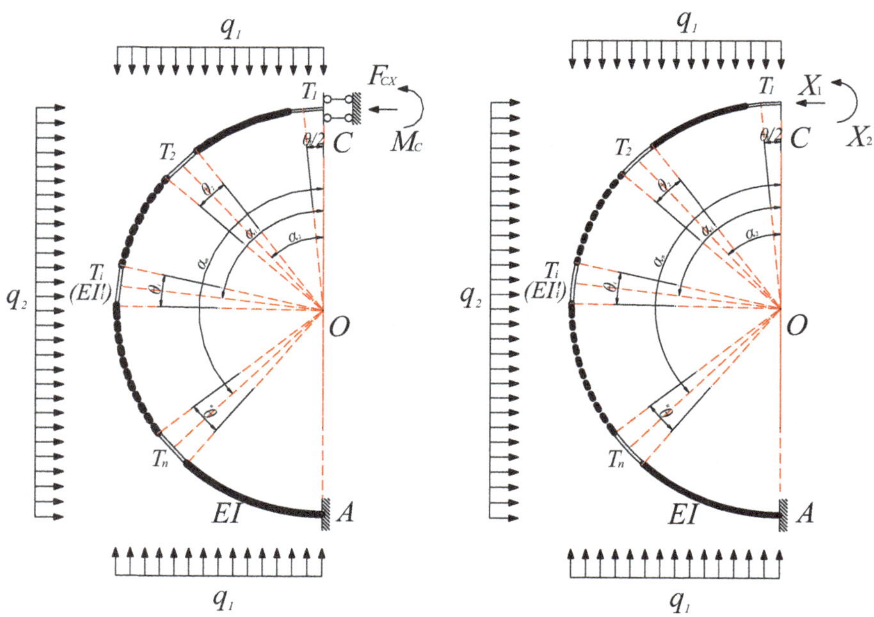

(a) Force analysis diagram (b) Force method basic system

Fig. 4.7 Mechanical model of circular arch

coal mine roadways usually have three to five sections. This section of this book finally establishes the arch mechanical models with arbitrary sections as shown in Fig. 4.7. T_i represents the ith joint from vault to arch bottom. The joint position is represented by α_i (joint center cross-section position). The joint flexural rigidity is $EI_i{}'$, and the joint length dimension is represented by θ_i. For the mechanical model shown in Fig. 4.7, the value of θ_1 is zero when there is no joint on the vault.

Symbol conventions: In the following mechanical calculation and analysis, the axial force is positive as the component under compression and negative as the component under tension. The bending moment is positive as the compression on the upper or outer surface of the component and negative as the tension on the upper or outer surface of the component. The load and reaction force are positive in the direction indicated by the arrow in the figure.

4.4.1.2 Solution

The force method is used to solve the quadratic statically indeterminate structure in Fig. 4.7. The constraint on the support at point C of the vault is replaced by unknown redundant forces X_1 and X_2.

The force method formula:

$$\begin{aligned}\delta_{11}X_1 + \delta_{12}X_2 + \Delta_{1P} = 0\\ \delta_{21}X_1 + \delta_{22}X_2 + \Delta_{2P} = 0\end{aligned} \tag{4.13}$$

In the above formula,

δ_{ij} is represents the generalized displacement generated at point i on the basic structure where the element j acting alone with unit force; Δ_{ip} is represents the generalized displacement generated at point i on the basic structure where the external load acting alone.

According to the principle of virtual work, the formula can be obtained:

$$\begin{aligned}\delta_{11} &= \int \frac{\overline{M_1}\,\overline{M_1}}{EI}ds + \int \frac{\overline{F_1}\,\overline{F_1}}{EA}ds\\
&= \int_{\alpha_1-\theta_1/2}^{\alpha_1+\theta_1/2} \frac{\overline{M_1}\,\overline{M_1}}{EI'}Rd\varphi + \int_{\alpha_1+\theta_1/2}^{\alpha_2-\theta_2/2} \frac{\overline{M_1}\,\overline{M_1}}{EI}Rd\varphi + \cdots + \int_{\alpha_i-\theta_i/2}^{\alpha_i+\theta_i/2} \frac{\overline{M_1}\,\overline{M_1}}{EI'}Rd\varphi\\
&+ \int_{\alpha_i+\theta_i/2}^{\alpha_{i+1}-\theta_{i+1}/2} \frac{\overline{M_1}\,\overline{M_1}}{EI}Rd\varphi + \cdots \int_{\alpha_{n-1}+\theta_{n-1}/2}^{\alpha_n-\theta_n/2} \frac{\overline{M_1}\,\overline{M_1}}{EI}Rd\varphi + \int_{\alpha_n-\theta_n/2}^{\alpha_n+\theta_n/2} \frac{\overline{M_1}\,\overline{M_1}}{EI'}Rd\varphi\\
&+ \int_{\alpha_1-\theta_1/2}^{\alpha_1+\theta_1/2} \frac{\overline{F_1}\,\overline{F_1}}{EA'}Rd\varphi + \int_{\alpha_1+\theta_1/2}^{\alpha_2-\theta_2/2} \frac{\overline{F_1}\,\overline{F_1}}{EA}Rd\varphi + \cdots + \int_{\alpha_i-\theta_i/2}^{\alpha_i+\theta_i/2} \frac{\overline{F_1}\,\overline{F_1}}{EA'}Rd\varphi\end{aligned}$$

$$+\int_{\alpha_i+\theta_i/2}^{\alpha_{i+1}-\theta_{i+1}/2}\frac{\overline{F_1}\,\overline{F_1}}{EA}Rd\varphi+\cdots+\int_{\alpha_{n-1}+\theta_{n-1}/2}^{\alpha_n-\theta_n/2}\frac{\overline{F_1}\,\overline{F_1}}{EA}Rd\varphi+\int_{\alpha_n-\theta_n/2}^{\alpha_n+\theta_n/2}\frac{\overline{F_1}\,\overline{F_1}}{EA'}Rd\varphi$$

$$=\sum_1^n\left(\int_{\alpha_i-\theta_i/2}^{\alpha_i+\theta_i/2}\frac{\overline{M_1}\,\overline{M_1}}{EI_i'}Rd\varphi+\int_{\alpha_i+\theta_i/2}^{\alpha_{i+1}-\theta_{i+1}/2}\frac{\overline{M_1}\,\overline{M_1}}{EI}Rd\varphi\right)$$

$$+\sum_1^n\left(\int_{\alpha_i-\theta_i/2}^{\alpha_i+\theta_i/2}\frac{\overline{F_1}\,\overline{F_1}}{EA'}Rd\varphi+\int_{\alpha_i+\theta_i/2}^{\alpha_{i+1}-\theta_{i+1}/2}\frac{\overline{F_1}\,\overline{F_1}}{EA}Rd\varphi\right) \tag{4.14}$$

Similarly, δ_{22}, δ_{12}, Δ_{1p}, and Δ_{2p} can be obtained, and the unknown redundant forces X_1 and X_2 can be obtained. Then, the moment and axial force of the arbitrary cross-section of a circular arch can be obtained according to formula (4.15).

$$\begin{aligned} M &= \overline{M}_1X_1+\overline{M}_2X_2+M_P \\ F &= \overline{F}_1X_1+\overline{F}_2X_2+F_P \end{aligned} \tag{4.15}$$

4.4.1.3 Example Analysis

Example 2 The radius R of the circular arch is 9.7 m; joints are set at 10°, 30°, 50°, 82°, 164°, 171° and 180°, and the effective length of the joint is 0.6 m; the bending rigidity *EI* of the cross section is 3300 kN m^2; the horizontal load q_2 born on the arch is 50 kN; and the lateral pressure coefficient of the initial value $\lambda = q_2/q_1 = 0.5$, $\mu = EI'/EI = 1.5$.

Figure 4.8 shows that under the load, the positive bending moment occurs at the vault and the arch bottom, indicating the inside of the arch is under tension; and the negative bending moment occurs at the 90° position, indicating the outside is under tension. The axial force is all pressure which are consistent with the force law of the arches in practical engineering. The force law indicates the internal force diagram is clearly symmetrical, and it is symmetrical up and down, and left and right.

4.4.1.4 Arch Force Law

The basic parameters of the circular arch in Example 2 are used to study the influence of load q_1, lateral pressure coefficient λ, rigidity ratio μ of casing joint, component rigidity *EI* and joint orientation angle α on the internal force of the arch.

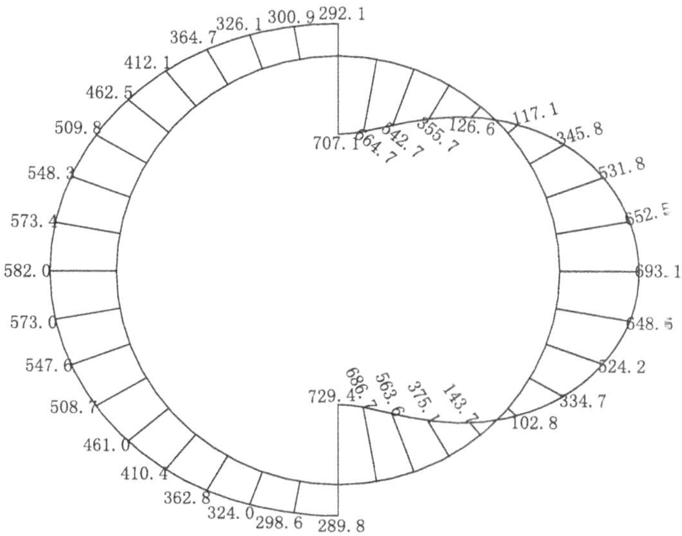

Fig. 4.8 Internal force calculation result of the circular arch

1. Influence of Load q_1 on Internal Force

With the other conditions of Example 2 unchanged, the diagrams of axial force and bending moment of the arch are drawn with different values of the load q_l. The analysis is made on the stress distribution of the arch. The results are shown in Fig. 4.9.

The analysis on Fig. 4.9 shows:

(1) The insides of both the vault and the arch bottom are under tension, and the outsides at the 90° position are under tension. This is consistent with the arch deformation law in practical engineering.
(2) The bending moment of the arch generally changes from the reverse to the positive near 40°, and changes back near 140°.
(3) As the load increases linearly, the internal force of the cross-section also increases linearly. The change of the load q_l will have different effects on the increase rate of internal forces in each cross-section.
(4) The changes of q_l does not affect the shape of the internal force diagram, but only changes the relative magnitude of the internal force.
(5) The maximum bending moment occurs at 180°, and the bending moment is also greater near 95°, while the maximum axial force appears near 90°.
(6) The axial force and the bending moment diagrams of the circular arch are symmetrical, and they are symmetrical left and right, and up and down.

2. Influence of Lateral Pressure Coefficient λ on Internal Force of the Arch

With the other conditions of Example 2 unchanged, an analysis is made on the force law of the arch with different lateral pressure coefficient λ, as shown in Fig. 4.10.

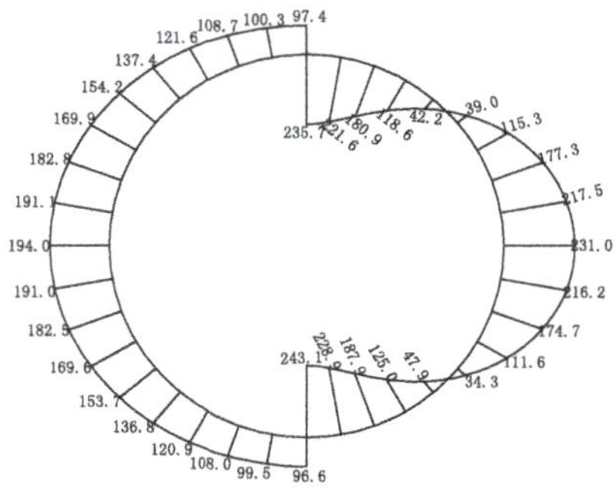

(a) q_1 =10kN/m

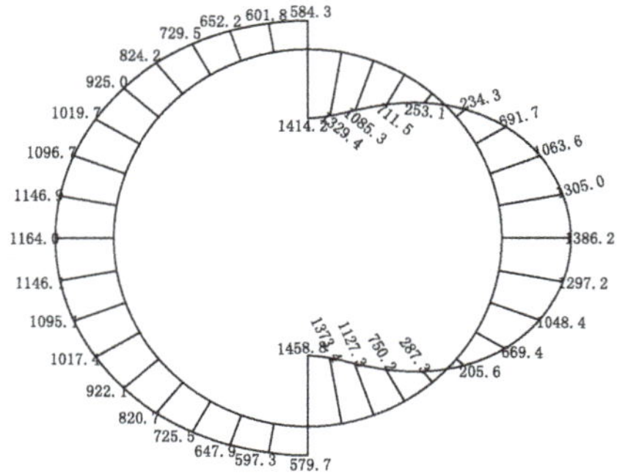

(b) q_1 =60kN/m

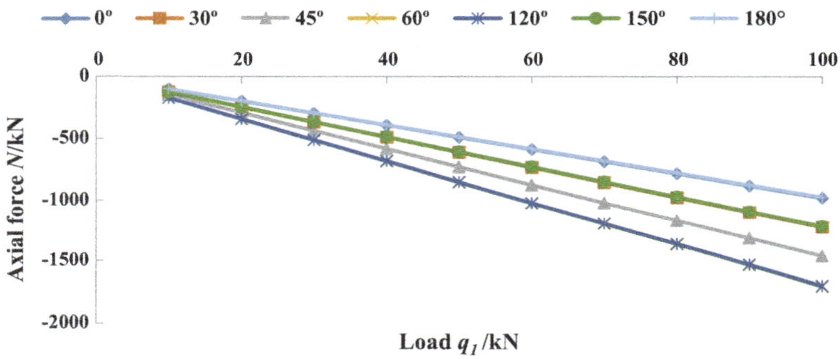

(c) Curves of axial force along with the changes of q_1

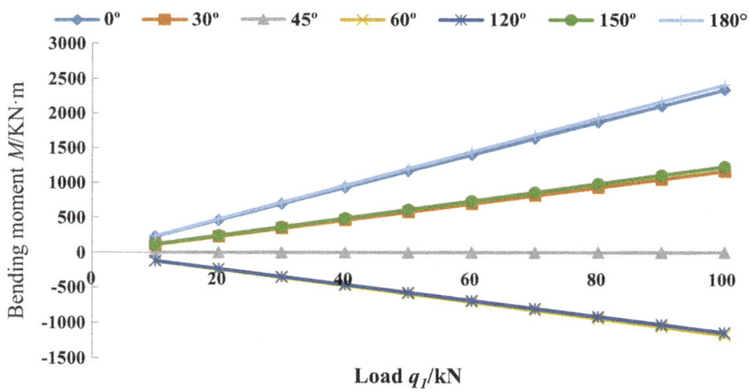

(d) Curves of bending moment along with the changes of q_1

Fig. 4.9 Curves of internal forces along with the changes of q_1 in different cross-sectional positions

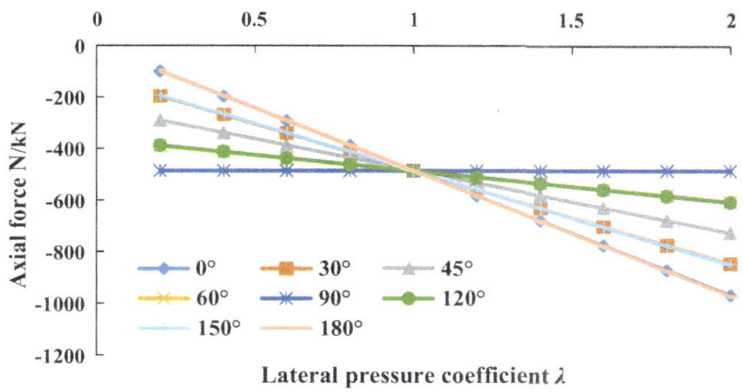

(a) Curves of axial force along with the change of λ

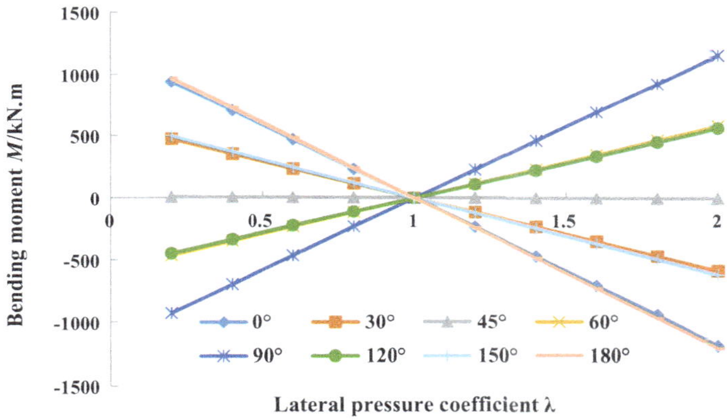

(b) Curves of bending moment along with the change of λ

Fig. 4.10 Curves of internal forces along with the change of λ in different cross-sectional positions

The analysis on Fig. 4.10 shows:

(1) The smaller the lateral pressure coefficient λ is, the inside of the vault and the outside at the 90° position more tend to be under tension. No matter what the value of the lateral pressure coefficient λ is, the inside of the arch bottom is under tension.
(2) When the lateral pressure coefficient λ is less than 1, the inside of the vault and the arch bottom are under tension, and the outside at the 90° position is under tension. When the lateral pressure coefficient λ is equal to 1, there is no bending moment on the cross-section of the arch. When the lateral pressure coefficient λ is more than 1, the outsides of the vault and the arch bottom are under tension, and the insides of the two arch sides are under tension.
(3) The closer the lateral pressure coefficient λ is to 1, the smaller the bending moment is on the cross-section of the arch.

(4) Within the variation arrange of the existing lateral pressure coefficient, the axial force of the arch is always the pressure. As the lateral pressure coefficient λ is less than 1, the closer λ is to 0; the smaller the axial force is on the vault and the arch bottom; and the larger the axial force is at the 90° position. As λ is more than 1, the farther away λ is from 1; the smaller the load is the 90° position; and the larger the load is on the vault. This is consistent with the equilibrium force state of the overall structure.
(5) The bending moment and axial force of each cross-section are linearly related to the lateral pressure coefficient λ.

3. The Influence of Equivalent Joint Rigidity Ratio μ on the Internal Force of the Arch

With the other conditions of Example 2 unchanged, the force law of the arch is analyzed with joint different rigidity ratios μ, as shown in Fig. 4.11.

Figure 4.11 shows:

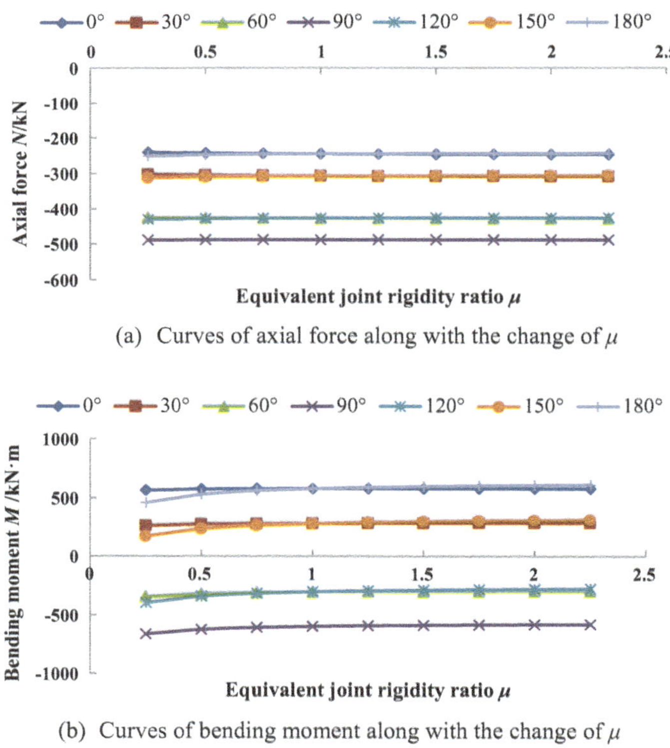

Fig. 4.11 Curves of internal force along with the change of μ in different cross-sectional positions

(1) The equivalent joint rigidity ratio μ has an influence on the internal force distribution of each cross-section. The influence is more significant on the bending moment.
(2) The variation of the equivalent joint rigidity ratio within the same magnitude will not fundamentally change the internal force of the arch cross-section.
(3) With the increase of μ, the bending moment is gradually increased on the tension position inside the arch and gradually decreased on the tension position outside the arch.
(4) The 90° cross-section is taken as the boundary. As the cross-section angle is less than 90°, the axial force decreases with the increase of μ, and vice versa. There is no change at the 90° cross-section, and in general, the variation range of overall arch is small.

4. Influence of Flexural Rigidity *EI* on Internal Force of Arch

With the other conditions of Example 2 unchanged, the force law on the arch is analyzed with different flexural rigidity *EI*, as shown in Fig. 4.12.

The analysis of Fig. 4.12 shows as the equivalent joint rigidity ratio μ is constant. The internal forces of each cross-section are not affected by the change of *EI*.

5. The Influence of the Joint Orientation Angle α on the Internal Force of the Arch

With the other conditions of Example 2 unchanged, the force law of the arch is analyzed as the joint changes from 0° to 30°, with removing 10°, 50°, and 171° joints and keeping the relative positions of 30° and 82° joint unchanged. The axial force and bending moment diagrams of the arch are drawn with different orientation angle α. The result is shown in Fig. 4.13.

The analysis of Fig. 4.13 shows with the other conditions unchanged the internal force of the arches has the following rules along with the change of α:

(1) The influence of the joint orientation angle is greater on the bending moment of the arch than on the axial force. The influence on the axial force is small.
(2) In general, the influence rate of the joint orientation angle is below 10% and mostly around 1% on the bending moment.
(3) The essence of studying the change of the joint orientation angle is to analyze the influence of the joint position change on the internal force. The cross-sectional bending moment will correspondingly vary with the joint orientation angle (Table 4.1).

6. Summary

Through the analysis of Table 4.1, the main influence of each factor on the internal force of a circular arch can be summarized as follows:

(1) The changes of the load and the lateral pressure coefficient have significant influences on the internal force. The changes of the equivalent joint rigidity ratio and the joint orientation angle have a smaller effect on the internal force. The flexural rigidity has no effect on the internal force.

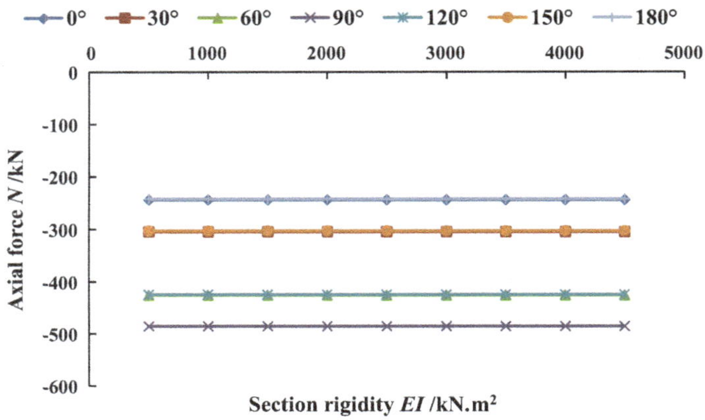

(a) Curves of axial force along with the change of *EI*

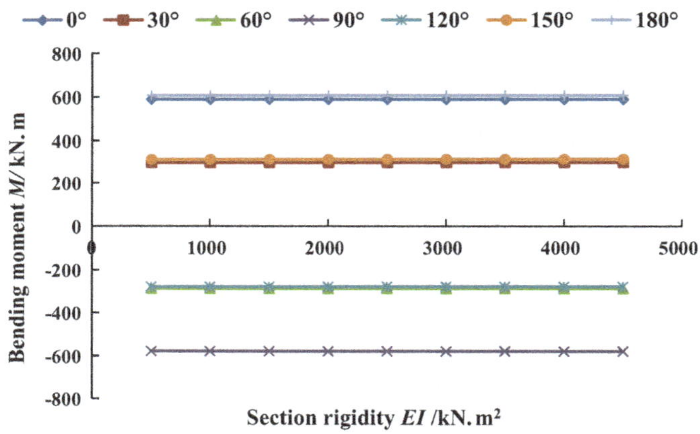

(b) Curves of bending moment along with the change of *EI*

Fig. 4.12 Curves of internal forces along with the change of *EI* at different cross-sectional positions

(2) The change of the load has a greater influence on the bending moment at 0° and 180° positions, and has no effect at 45° and 135° positions with zero bending moment near these points.
(3) The change of the load has a significant influence on the axial force at 90° position, and it is an extreme point.
(4) As the lateral pressure coefficient is 1, there is only axial force in each cross-section of the circular arch without bending moment.
(5) The change of the lateral pressure coefficient has no effect on the bending moment at 45° and 135° positions which is 0. It has the greatest influence at the 90° position.

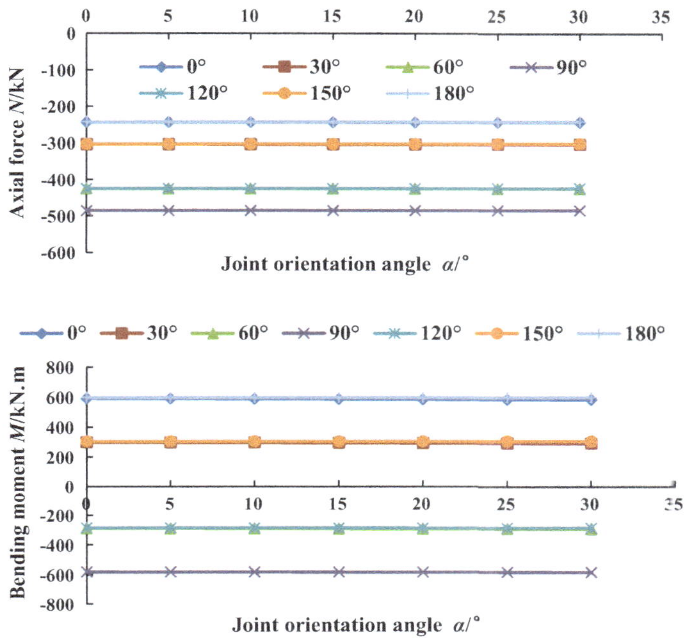

Fig. 4.13 Curves of the internal forces along with the change of α in different cross-sectional positions

4.4.2 *Internal Force Calculation of Straight-Wall Semi-circular Arches*

4.4.2.1 Calculation Methods and Mechanical Model

The mechanical models of the arches with arbitrary sections are shown in Fig. 4.14. T_i represents the *i*th casing from the vault to the arch bottom with its position represented by α_i (the position of the joint center cross-section), the flexural rigidity by EI_i' and the casing length by θ_i. For the mechanical model shown in Fig. 4.14, $\theta_1 = 0$ as the number of arches sections is odd, and $\theta_1 \neq 0$ as the number of arches sections is even.

4.4.2.2 Solution

1. Solution for Support Reaction Force

Figure 4.15a shows the internal force diagram of the structure and the equilibrium formula of the internal force of the structure can be obtained:

Table 4.1 Statistics of the influence of each factor on the internal force of a circular arch

	Load q_1	Lateral pressure coefficient λ	Equivalent joint rigidity ratio μ	Flexural rigidity EI	Joint orientation angle α
Moment M is positive or negative, symmetrical up and down, and left and right	Positive linear correlation, having the greatest influence at 0° and 180° positions and no influence at 45° and 135° positions	Linear correlation; no effect when $\lambda = 1$; no effect at 45° and 135° positions; most significant effect at 90° position; and the effect is symmetrical with 90° position as the boundary (such as 60° and 120° positions)	Weak positive correlation	No effect	Basically no effect
The axial force N is all negative, and symmetrical up and down, and left and right	Linear positive correlation; 90° position is the extreme point; and the farther away from 90° position, the smaller effect is	Positive linear correlation; no effect at 90° position, farther away from 90° position, the more obvious the effect is, and the effect is symmetrical (such as 60° and 120° positions)	Weak negative correlation	No effect	Basically no effect
Internal force diagram is symmetrical up and down, and left and right	Only affecting the size of internal forces, not affecting the internal force diagrams	A great influence on the internal force and internal force diagram	Basically no effect	No effect	Basically no effect
Special angle	Zero moment points near 45° and 135° positions	Equal moment moments at 35° and 150° positions, and equal axial forces at 90° position			

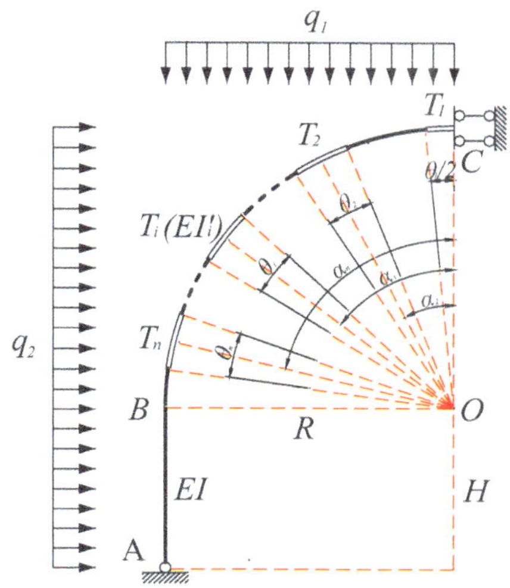

Fig. 4.14 Schematic diagram of the calculation of the mechanical model of the arch with arbitrary sections

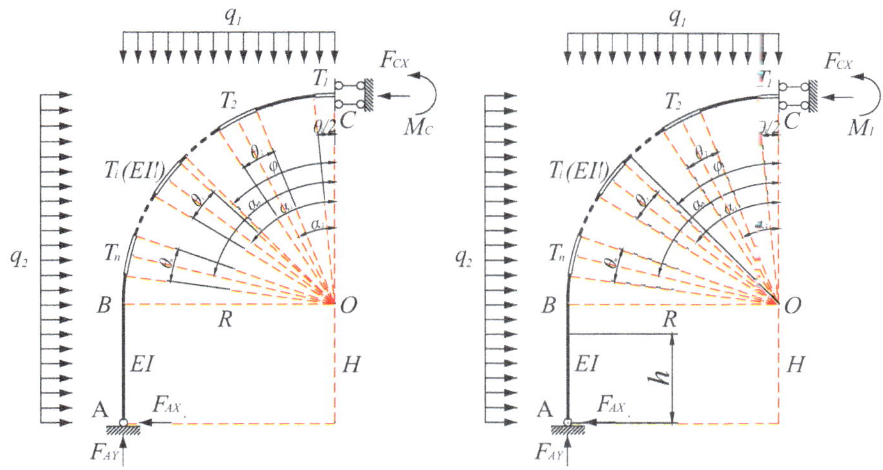

(a) External force and support reaction (b) Force method basic system

Fig. 4.15 Calculation of internal force of arch

$$\begin{cases} F_{AY} - q_1 R = 0 \\ F_{AX} + F_{CX} - q_2(R + H) = 0 \\ F_{AX}(H + R) + F_{AY} R - \frac{1}{2} q_2 (H + R)^2 - \frac{1}{2} q_1 R^2 - M_C = 0 \end{cases} \tag{4.16}$$

F_{AX}, F_{AY}, F_{CY} and M_C are the support reaction forces. q_1 and q_2 are the simplified forces of the surrounding rock on the arches. R is the centerline radius of the arches and H is the straight leg height of the arches.

The force method is used to solve the onetime statically indeterminate structure.

With M_c as the redundant unknown force, the basic system of force method is shown in Fig. 4.15b, and the basic unknown force is M_1.

The force formula is: $\delta_{11}X_1 + \Delta_{1P} = 0$.

The known support reactions can be solved by bringing the obtained M_C into the primary force equilibrium formula (4.16).

2. Internal Force Solution

As each support reaction is known, the internal force of the cross-section with the angle φ can be solved between the support and the center cross-section of the casing joint:

Axial force:

$$F_{N\varphi} = F_{CX}\cos\varphi + q_1 R\sin^2\varphi - q_2 R\cos\varphi(1-\cos\varphi) \tag{4.17}$$

$$F_{Nh} = F_{AY} = q_1 R \tag{4.18}$$

Bending moment:

$$M_\varphi = M_C + F_{CX}R(1-\cos\varphi) - \frac{1}{2}q_1R^2\sin^2\varphi - \frac{1}{2}q_2R^2(1-\cos\varphi)^2 \tag{4.19}$$

$$M_h = F_{AX}h - \frac{1}{2}q_2h^2. \tag{4.20}$$

4.4.2.3 Example Analysis

Example 3 A straight-wall semi-circular arch has four sections ($n = 2, \theta_I \neq 0$), and its axis radius $R = 2500$ mm. The straight leg height $H = 2000$ mm. One casing is located at vault and another casing is located at $\alpha_2 = 55°$ position. The radian is the same at each joint $\theta = 20°$. The bearing load of the arches $q_I = 0.1$ MPa and the lateral pressure coefficient $\lambda = 1.5$. The flexural rigidity of confined concrete components $EI = 3300$ kN m^2. The casing joints are in a rigid joint state and the effective flexural rigidity ratio of each casing joint is the same, $\mu = 1.5$, that is $EI' = 1.5\,EI = 4950\,\text{kN m}^2$. With all the known conditions, each support reaction and the internal force of the arch can be solved.

Bring the known conditions into formula (4.16), following results can be obtained through calculations:

$$F_{AX} = 236.4\,\text{kN},\ F_{AY} = 250.0\,\text{kN},\ F_{CX} = 435.6\,\text{kN},\ M_C = -129.0\,\text{kN m}.$$

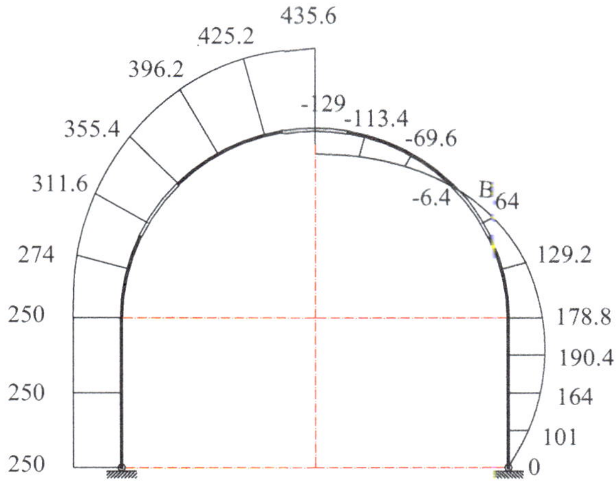

Fig. 4.16 Calculation result of internal force of arch

The calculation results of the axial force and the bending moment are drawn in Fig. 4.16. The left half of the figure is the axial force diagram in kN, and the right half is the bending moment diagram in kN m.

The axial force and the bending moment of the arches show:

(1) The maximum value of the axial force is 435.6 kN, which is located at the vault. The lower the position is, the lower the axial force is. The minimum value is 250 kN at the skewbacks. The axial force of the straight legs is equal everywhere.
(2) The bending moment as a whole shows a positive value at the lower part and negative value at the upper part. The moment 0 is located near the 45° position above the skewback. The maximum value of the negative moment is −129 kN m at the vault and the maximum value of the positive moment is approximately 190.4 kN m near 3/4 height of the straight leg.

4.4.2.4 Arch Force Law

The basic parameters of the circular arch of Example 3 are used to study the influence law of load q_1, lateral pressure coefficient λ, arch flexural rigidity EI, equivalent joint rigidity ratio μ, joint orientation angle α, and height-diameter ratio κ on internal force of arches.

The axial force and bending moment diagrams of the arch are drawn through the analysis method of the influence law of the internal force of the circular arch in Sect. 4.4.1.4 with different value of each parameter. The force of the arch is analyzed, and the results are shown in Table 4.2.

Through the above analysis, the main influence of each factor can be summarized on the internal force of straight-wall semi-circular arches:

Table 4.2 The influence of various factors on the internal force of straight-wall semi-circle arch

	Load q_1	Lateral pressure coefficient λ	Equivalent joint rigidity ratio μ	Flexural rigidity EI	Joint orientation angle α	High-diameter ratio κ
Bending moment M is positive or negative, and symmetrical left and right	Absolute value of bending moment has a positive linear correlation with it	Linear; bending moment changes its positive and negative sign with the increase of the lateral pressure coefficient	Weak positive correlation	No effect	Weak negative correlation	The absolute value of the bending moment has a positive correlation with it
The axial force N is all positive, and symmetrical left and right	Positive linear correlation	Positive linear correlation; the closer to the vault is, the more obvious the influence is; no effect on the straight leg	Weak negative correlation; the closer to the straight leg is, the smaller the influence is, and no effect on the straight leg	No effect	Weak positive correlation; the closer to the straight leg is, the smaller the influence is, and no effect on the straight leg	The exponential positive correlation; the closer to the vault is, the more obvious the influence is, and no effect on the straight leg
Internal force diagram	Only affect the size, Not affecting the shape	Affecting both size and shape	Basically no effect	No effect	Basically no effect	Changing not only the size but also the shape; but not as significant as λ

(1) Axial force is all positive, and bending moments are both positive and negative. Axial force and bending moment diagrams are symmetrical left and right;
(2) The most significant influence factors on the internal force are load q_l, lateral pressure coefficient λ, and height-diameter ratio κ;
(3) Axial force and bending moment (absolute value) are positively related to load q_l and height-diameter ratio κ;
(4) The load q_1 only affects the size of the internal force and does not affect the shape of the internal force diagram. The lateral pressure coefficient λ and the height-diameter ratio κ affect both the internal force and the internal force diagram;
(5) The equivalent joint rigidity ratio μ and the joint orientation angle α have little effect on the internal force;
(6) The flexural rigidity EI had no effect on the internal force of the arch.

4.5 Analysis of Bearing Capacity of the Arch

After the internal force of the arch is obtained, the ultimate bearing capacity of the components of SQCC, CCC and U-steel is analyzed through combining with the calculation formulas of ultimate bearing capacity.

4.5.1 The Ultimate Bearing Capacity of SQCC Components

The calculation results of the internal force of the arches show that the acting force of the arch is mainly in the bending state, and therefore the bending strength criterion of the components needs to be obtained.

1. Calculation Formula

The bearing capacity of SQCC components can be calculated according to the following formula.

As $N/N_u \geq 2\eta_0$,

$$\frac{N}{N_u} + \frac{a \cdot \beta_m \cdot M}{M_u} \leq 1 \tag{4.21}$$

As $N/N_u < 2\eta_0$

$$\frac{-b \cdot N^2}{N_u^2} - \frac{c \cdot N}{N_u} + \frac{\beta_m \cdot M}{M_u} \leq 1 \tag{4.22}$$

In the formula, N is axial pressure, M is a bending moment, N_u is the ultimate axial force, M_u is the ultimate bending moment, β_m is the equivalent moment coefficient (value is 1.0), ξ is the confinable effect coefficient, f_y is the steel yielding strength, f_{cu} is the compressive strength of concrete cubes, f_{sc} is the compressive strength of the steel pipe filled with concrete, α is the steel ratio of the cross-section of the steel pipe component, γ_m is the calculation coefficient of the ultimate bending moment, A_{sc} is the cross-section area of the steel pipe, and W_{scm} is the bending modulus of component cross-section, η_0, ζ is the coefficient related to ξ, and a, b and c are the coefficients relating to η_0, ζ.

2. Example Analysis

Example 4 The component model is SQCC 150 × 8 with edge length of 150 mm, the wall thickness of 8 mm, and C40 concrete filled inside. With the known conditions, the calculation formula of the ultimate bending moment will be deduced.

Material parameters are substituted into the formula:

$$f_{ck} = 26.8\,\text{MPa},\ f_y = 409\,\text{MPa},\ \alpha_s = 0.25,\ \xi = \frac{\alpha_s f_y}{f_{ck}} = \frac{0.25 \times 409}{26.8} = 3.66$$
$$A_{sc} = 0.15^2 = 0.0225\,\text{m}^2,\ \gamma_m = 1.04 + 0.48\ln(\xi + 0.1) = 1.724,$$
$$W_{scm} = B^3/6 = 0.00056\,\text{m}^3,$$
$$f_{sc} = (1.18 + 0.85\xi) f_{ck} = 115.0\,\text{MPa},\ N_u = f_{sc} \cdot A_{sc} = 2.587 \times 10^6\,\text{N},$$
$$M_u = \gamma_m W_{scm} f_{sc} = 111.0 \times 10^3\,\text{N m}$$
$$\zeta = 1 + 0.14\xi^{-1.3} = 1.0259,\ \eta_0 = 0.1 + 0.13 \cdot \xi^{-0.81} = 0.1454$$
$$a = 1 - 2 \cdot \eta_0 = 0.7091,\ b = \frac{1-\zeta}{\eta_0^2} = -1.2251,\ c = \frac{2 \cdot (\zeta - 1)}{\eta_0} = 0.356$$

Set $n = N/N_u$, $m = M/M_u$

The calculation formula can be obtained for the ultimate bending moment of SQCC150 × 8 (C40) components:

$$\begin{cases} 1.225n^2 - 0.3563n + m - 1 \le 0\ , & n < 0.2908 \\ n + 0.7091m - 1 \le 0, & n \ge 0.2908 \end{cases} \tag{4.23}$$

According to the above information, the corresponding m–n curve can be drawn as shown in Fig. 4.17.

Specific physical meaning of m–n curve: After the axial force N and the bending moment M of the component are obtained, m and n are obtained through further calculations. When the corresponding positions of m and n in the figure are within the envelope of the indicated curve and the positive direction of the coordinate axis, the strength of the component is reliable. There is no strength failure of the component. On the contrary, strength failure occurs.

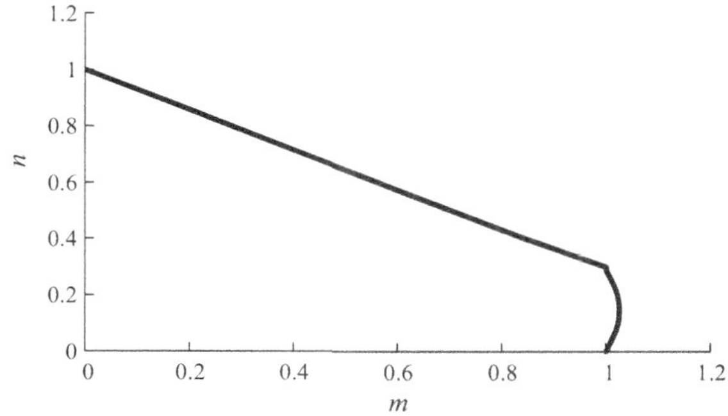

Fig. 4.17 The *m*–*n* curve of the SQCC150 × 8 (C40) components

4.5.2 Ultimate Bending Moment of CCC Component

1. Calculation Formula

The ultimate bending moment of the circular confined concrete component can be calculated according to the following formula:

When $N/N_u \geq 2\eta_0$

$$\frac{N}{N_u} + \frac{a \cdot \beta_m \cdot M}{M_u} \leq 1 \tag{4.24}$$

When $N/N_u < 2\eta_0$

$$\frac{-b \cdot N^2}{N_u^2} - \frac{c \cdot N}{N_u} + \frac{\beta_m \cdot M}{M_u} \leq 1 \tag{4.25}$$

In the formula, N is axial pressure; M is the maximum designed value of the bending moment in the calculating range; N_u is the ultimate axial force; M_u is ultimate bending moment; β_m is equivalent moment coefficient (taken 1.0); ξ *is* confinable effect coefficient; f_y is steel yielding strength; f_{cu} is the compressive strength of concrete cubes; f_{sc} is the compressive strength of confined concrete; α is the cross-sectional steel ratio of the confined concrete component γ_m is the calculation coefficient of the ultimate bending moment; A_{sc} is the cross-sectional area of confined concrete; W_{scm} is the bending modulus of the component cross-section; η_0, ζ are the coefficients related to ξ; and a, b, and c are the coefficients relating to η_0 and ζ.

2. Example Analysis

Example 5 Component model is CCC159 × 8 with the outside diameter of 159 mm, the wall thickness of 8 mm and C40 concrete filled inside. The calculation formula is deduced for the ultimate bending moment with those parameters.

Material parameters are substituted into the formula:

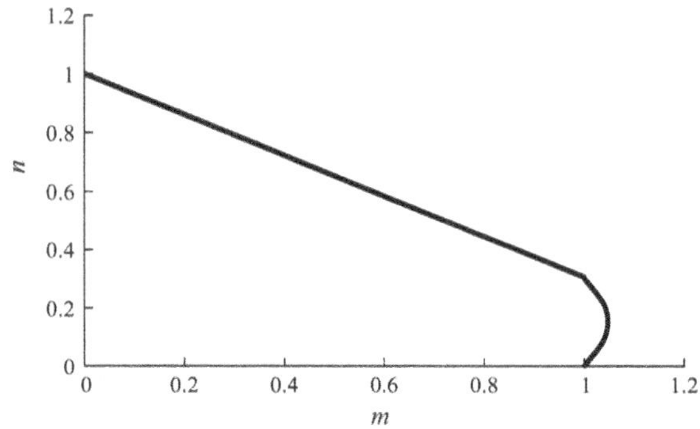

Fig. 4.18 *m–n* curve of CCC159 × 8 (C40) components

$$f_{ck} = 26.8\,\text{MPa},\ f_y = 345\,\text{MPa},\ \alpha_s = 0.24, \xi = \frac{\alpha_s f_y}{f_{ck}} = \frac{0.24 \times 345}{26.8} = 3.09$$

$$A_{sc} = \frac{\pi(0.159)^2}{4} = 0.0199\,\text{m}^2,\ \gamma_m = 1.1 + 0.48\ln(\xi + 0.1) = 1.676,$$

$$W_{scm} = \frac{\pi \cdot (0.159)^3}{32} = 0.00039\,\text{m}^3,$$

$$f_{sc} = (1.14 + 1.02\xi) f_{ck} = 115.0\,\text{MPa},\ N_u = f_{sc} \cdot A_{sc} = 2.29 \times 10^6\,\text{N},$$

$$M_u = \gamma_m W_{scm} f_{sc} = 64.4 \times 10^3\,\text{N m}$$

$$\zeta = 0.18\xi^{-1.15} + 1 = 1.049,\ \eta_0 = 0.1 + 0.14 \cdot \xi^{-0.84} = 0.1527$$

$$a = 1 - 2 \cdot \eta_0 = 0.695,\ b = \frac{1 - \zeta}{\eta_0^2} = -2.101,\ c = \frac{2 \cdot (\zeta - 1)}{\eta_0} = 0.6418$$

The calculation formula can be obtained for the ultimate bending capacity of CCC159 × 8 (C40) components:

$$\begin{cases} 2.019n^2 - 0.6155n + m - 1 \le 0\ , & n < 0.3054 \\ n + 0.695m - 1 \le 0, & n \ge 0.3054 \end{cases} \tag{4.26}$$

The corresponding *m–n* curve is shown in Fig. 4.18.

4.5.3 Ultimate Bending Capacity of I-Steel Components

The cross-section of I-steel is made approximately equivalent to the ideal I-steel cross-section as shown in Fig. 4.19. When the axial force and bending moment work together on the cross-section, the internal forces can be obtained respectively under the limit state according to the stress distribution (as shown in Figs. 4.20 and 4.21).

When only the axial force acts on the cross-section, its ultimate axial force is:

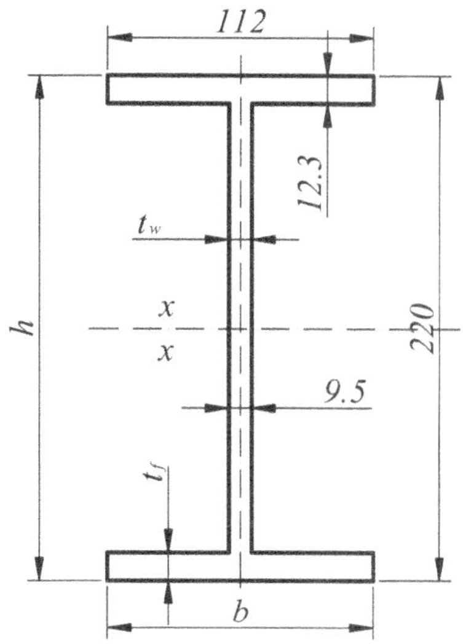

Fig. 4.19 The ideal cross-section of I22b I-steel (mm)

$$N_u = A \cdot \sigma_s \tag{4.27}$$

When the bending moment alone acts on the cross-section, its ultimate bending moment is (the fully plastic state shown in Fig. 4.20c):

$$M_u = t_w(0.5h - t_f)^2 \cdot \sigma_s + bt_f(h - t_f) \cdot \sigma_s \tag{4.28}$$

where, N_u is ultimate axial force, M_u is ultimate bending moment; A is I-steel cross-sectional area; σ_s is ultimate yield stress of the material; I-steel cross-sectional parameters (ideal cross-section, see Fig. 4.19): h is height, b is width; t_w is web plate thickness; t_f is flange plate thickness (Fig. 4.20).

With the stress state of the cross-section under the compression-bending shown in Fig. 4.21, the internal force of the component can be calculated and dimensionless.

Set:

$$n = \frac{N}{N_u} \tag{4.29}$$

$$m = \frac{M}{M_u} \tag{4.30}$$

where, N—axis force and M—bending moment.

According to the stress state of the cross-section, the expressions are deduced for axial force N and bending moment M and brought into formulas (4.29) and (4.30). The following generalized yield criteria are obtained for I-steel cross-sections:

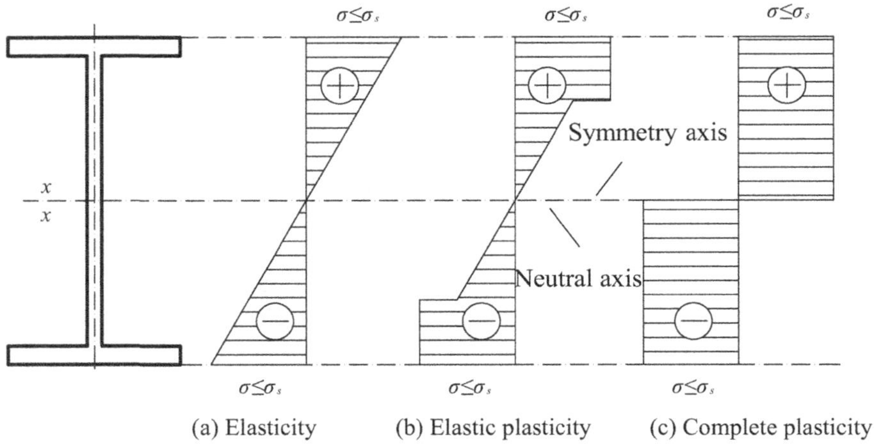

Fig. 4.20 The stress distribution of I-steel only by moment

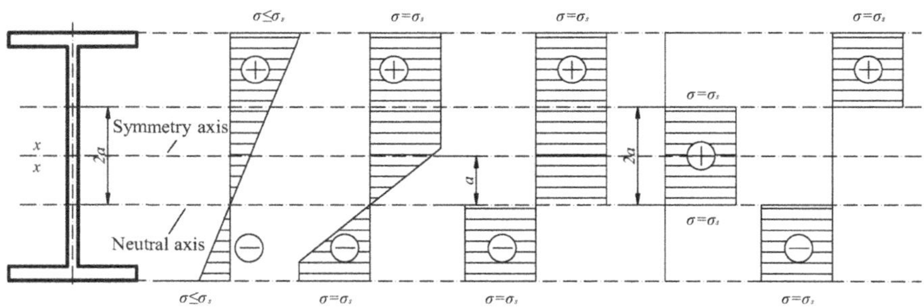

Fig. 4.21 The stress distribution of I-steel under the compression-bending

When $n \leq \frac{A_w}{A}$ (the neutral axis is in the web),

$$m + n^2 \left[\frac{1}{1 - (\frac{A_f}{A})^2 (1 - \frac{t_w}{b})} \right] = 1 \tag{4.31}$$

When $n > \frac{A_w}{A}$ (the neutral axis is in the flange),

$$m + \left[\frac{n^2 - (1 - \frac{t_w}{b})(n - \frac{A_w}{A})^2}{1 - (\frac{A_f}{A})^2 (1 - \frac{t_w}{b})} \right] = 1 \tag{4.32}$$

In the formula, A is cross-sectional area; h is height and b is width; t_w is web plate thickness; t_f is flange plate thickness; A_f is flange plate area; A_w is web plate area, A_a is axis force calculation area $A_a = \frac{N}{f_y}$.

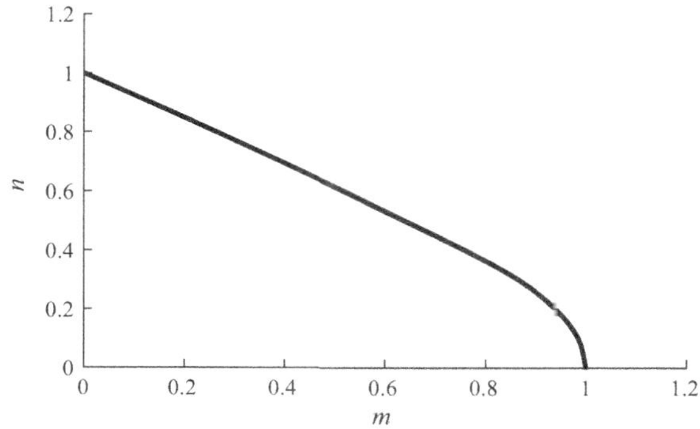

Fig. 4.22 *m–n* curve of I22b-steel

Example 6 The high-strength I22b-steel (36.4 kg/m) is commonly used for supports in underground engineering. The cross-section parameters are shown in Fig. 4.19. The generalized yield function is obtained by the above formulas.

When $n \leq 0.4$ (the neutral axis is in the web):

$$m + 1.485n^2 = 1 \tag{4.33}$$

When $n > 0.4$ (neutral axis is in the flange):

$$m + 0.126n^2 + 1.09n - 0.217 = 1 \tag{4.34}$$

The *m–n* curve is drawn in Fig. 4.22.
where, $n = \frac{N}{N_u}$, $m = \frac{M}{M_u}$,

$$\begin{aligned} N_u &= A \cdot \sigma_s = 46.1 \times 10^{-4} \times 235 \times 10^6 = 1083.7\,\text{kN}, \\ M_u &= t_w(0.5h - t_f)^2 \cdot \sigma_s + bt_f(h - t_f) \cdot \sigma_s \\ &= 71.3\,\text{kN m}. \end{aligned}$$

4.5.4 Ultimate Bending Capacity of U-Steel Component

The contour of the U-steel cross-section is irregular and has only one axis of symmetry. For analysis convenience, the U-steel cross-section can be simplified to be equivalent to I-steel cross-section which is asymmetric up and down, as shown in Fig. 4.23. The equivalent I-steel parameters of U-steel are given in Table 4.3.

The ultimate bearing capacity of I-steel components can be calculated according to the following formula:

When $n \leq \frac{A_w}{A}$ (the neutral axis is in the web),

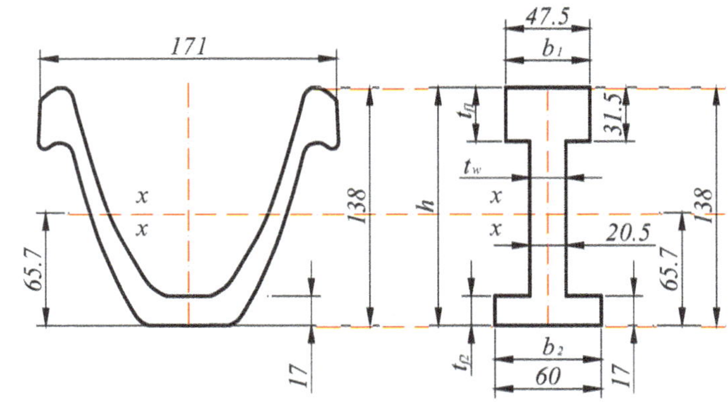

Fig. 4.23 The parameters of the U36 steel cross-section equivalent to I-steel cross-section (mm)

Table 4.3 I-steel cross-section coefficient of U-shape steel (mm)

	h	b_1	b_2	t_{f1}	t_{f2}	t_w	x
U18	100	38	58.5	18	10	12.9	49.5
U25	110	42	52.2	26	17	14.4	63.5
U29	121	40.8	54	28.5	16	20.1	66.3
U36	138	47.5	60	31.5	17	20.5	65.7

$$m + n^2\left[\frac{1}{1-\left(\frac{A_f}{A}\right)^2\left(1-\frac{t_w}{b}\right)}\right] = 1 \tag{4.35}$$

When $n > \frac{A_w}{A}$ (the neutral axis is in the flange),

$$m + \left[\frac{n^2-\left(1-\frac{t_w}{b}\right)\left(n-\frac{A_w}{A}\right)^2}{1-\left(\frac{A_f}{A}\right)^2\left(1-\frac{t_w}{b}\right)}\right] = 1 \tag{4.36}$$

where, $n = \frac{N}{N_u}$; $m = \frac{M}{M_u}$; N is axis force; M is bending moment; A is cross-section area of the I-steel; h is the height of the I-steel cross-section, b is the width of the I-steel cross-section; t_w is web plate thickness of I-steel; t_f is flange plate thickness of I-steel; A_f is flange area of I-steel; A_w is web area of I-steel; and A_a is axis force calculation area of I-steel $A_a = \frac{N}{f_y}$.

With the reference to the bearing capacity calculating method of I-steel cross-section, the ultimate bearing capacity of U36 steel can be calculated (Fig. 4.24).

Positive bending:

When $n \le 0.3017$ (the neutral axis is in the web),

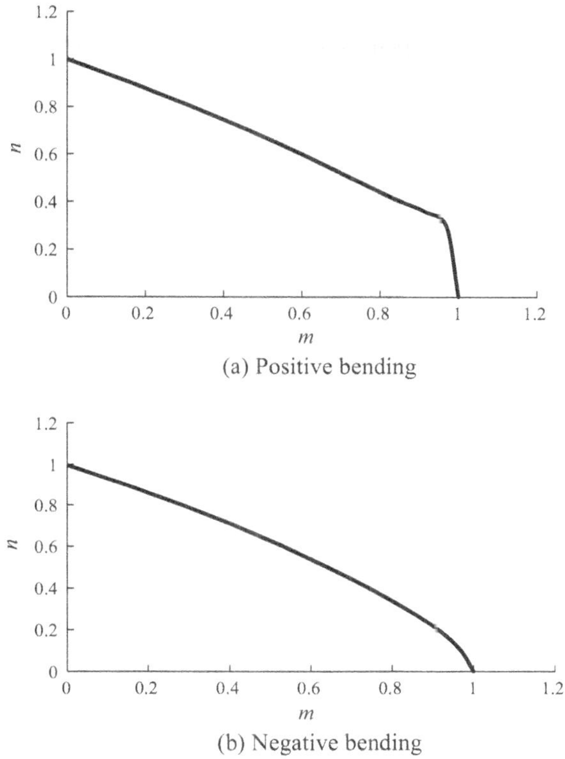

Fig. 4.24 Calculation of ultimate bending capacity of U36 steel

$$m = -0.717n^2 - 0.062n + 1 \tag{4.37}$$

When $n > 0.3017$ (the neutral axis is in the flange),

$$m = -0.51n^2 - 0.685n + 1.1951 \tag{4.38}$$

Negative bending:
When $n \leq 0.3122$ (the neutral axis is in the web),

$$m = -1.2525n^2 - 0.1810n + 1 \tag{4.39}$$

When $n > 0.3122$ (the neutral axis is in the flange),

$$m = -0.479n^2 - 0.5755n + 1.05 \tag{4.40}$$

where, $n = \frac{N}{N_u}$, $m = \frac{M}{M_u}$,

$$N_u = A \cdot \sigma_s = 43.5 \times 10^{-4} \times 335 \times 10^6 = 1457.6\,\text{kN},$$

Table 4.4 Statistics of bending bearing capacity q_{1cr} of various components

Types	Formulas	N_u/kN	M_u/kN
SQCC150 × 8	$1.225n^2 - 0.3563n + m - 1 \leq 0$, $n < 0.2908$ $n + 0.7091m - 1 \leq 0,\quad n \geq 0.2908$	2587	111.0
CCC159 × 8	$2.019n^2 - 0.6155n + m - 1 \leq 0$, $n < 0.3054$ $n + 0.695m - 1 \leq 0,\quad n \geq 0.3054$	2290	64.4
U36	Positive curve: $m + 0.717n^2 + 0.062n - 1 \leq 0, n \leq 0.3017$ $m + 0.51n^2 + 0.685n - 1.1951 \leq 0, n > 0.3017$ Negative curve: $m + 1.2525n^2 + 0.1810n - 1 \leq 0, n \leq 0.3122$ $m + 0.479n^2 + 0.5755n - 1.05 \leq 0, n > 0.3122$	1457.6	48.3

$$\begin{aligned} M_u &= t_{f1}b_1(h - x - 0.5t_{f1}) \cdot \sigma_s + 0.5t_w(h - x - t_{f1})^2 \cdot \sigma_s \\ &+ t_{f2}b_2(x - 0.5t_{f2}) \cdot \sigma_s + 0.5t_w(x - t_{f2})^2 \cdot \sigma_s \\ &= 48.3\,\text{kN m}. \end{aligned}$$

4.5.5 Comparative Analysis of Ultimate Bearing Capacity of Arches

1. Calculate Formula Statistics for Ultimate Bearing Capacity

The ultimate bearing capacity of CCC159 × 8, SQCC150 × 8 and U36 components with the same steel content is shown in Table 4.4.

2. Calculation of Ultimate Bearing Capacity of Arches with Different Cross-sections

Based on Table 4.4, a judgment and an analysis are made on the results of the internal force calculation in Sect. 4.4. The ultimate bearing capacity of the arches with different cross-sections and sections is obtained. The results are summarized in Tables 4.5 and 4.6, in which the dimensions of the arches are the same with those in Sect. 4.4.

The analysis of Tables 4.5 and 4.6 and Figs. 4.25, 4.26 and 4.27 shows:

(1) The lateral pressure coefficient has a significant influence on the ultimate bearing capacity of the arches. With the increase of the lateral pressure coefficient, the ultimate bearing capacity of the arches shows a trend of increasing first and then decreasing. The change is obvious.

Table 4.5 Statistics of ultimate bearing capacity q_{1cr} of straight-wall semi-circular arches with different cross-sections (MPa)

λ	0.25	0.5	0.75	1	1.25	1.5	2
SQCC150 × 8	0.169	0.389	0.191	0.112	0.079	0.06	0.041
CCC159 × 8	0.098	0.245	0.114	0.065	0.051	0.035	0.024
U36	0.071	0.155	0.078	0.046	0.033	0.025	0.018

Table 4.6 Statistics of ultimate bearing capacity q_{1cr} of circular arches with different cross-sections (MPa)

λ	0.25	0.5	0.75	1	1.25	1.5	2
SQCC150 × 8	0.096	0.146	0.287	1.05	0.284	0.146	0.073
CCC159 × 8	0.0554	0.085	0.172	0.912	0.174	0.086	0.042
U36	0.040	0.059	0.109	0.580	0.109	0.058	0.0298

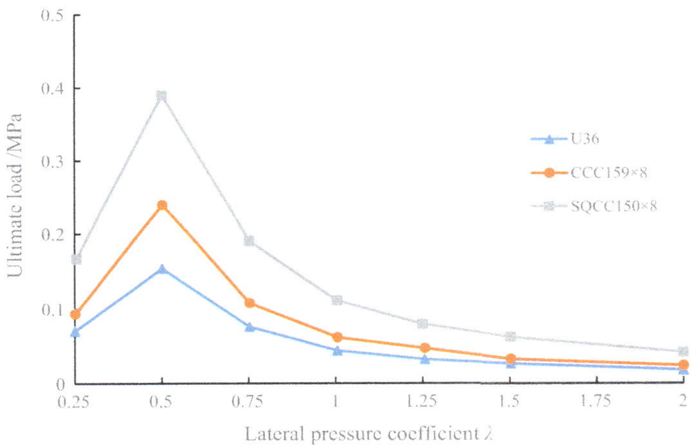

Fig. 4.25 Curves of the ultimate bearing capacity q_{1cr} of straight-wall semi-circular arches along with the changes of λ

(2) The ultimate bearing capacity is reached as the lateral pressure coefficient of the circular arch is 1. SQCC150 × 8 is taken as an example. As the lateral pressure coefficient of the circular arch is 1, the ultimate bearing capacity is 10.9 times and 14.4 times of that as the lateral pressure coefficient is 0.25 and 2.0.

(3) The maximum ultimate bearing capacity is reached as the lateral pressure coefficient of the straight-wall semi-circular arch is 0.5. SQCC150 × 8 is taken as an example. As the lateral pressure coefficient of the circular arch is 1, the ultimate bearing capacity is 2.3 times and 9.49 times of that as the lateral pressure coefficient of the circular arch is 0.25 and 2.0.

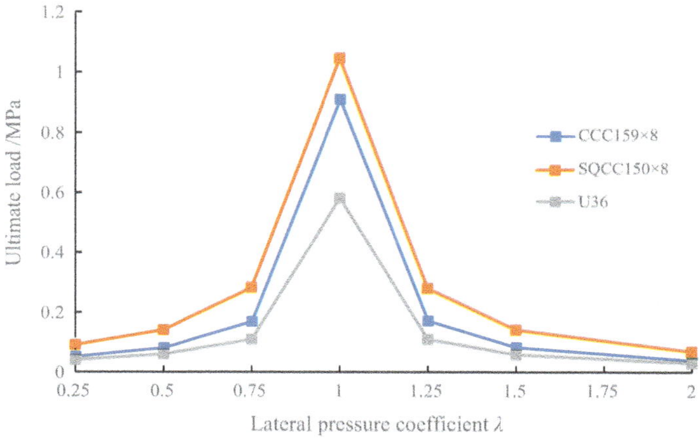

Fig. 4.26 Curves of the ultimate bearing capacity q_{1cr} of circular arches along with the changes of λ

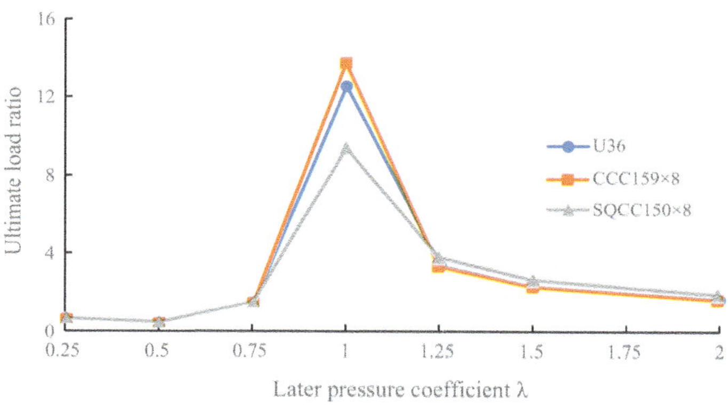

Fig. 4.27 Curve of the ratio of the ultimate bearing capacity q_{1cr} of circular/straight-wall semi-circular arches along with the changes of λ

(4) The comparison of the ultimate bearing capacity of arches with different cross-sections shows as the lateral pressure coefficient is 0.5, the maximum ultimate bearing capacity of the straight-wall semi-circular arch of SQCC150 × 8 is 0.389 MPa, which is 2.51 times that of U36. As the lateral pressure coefficient is 1, the maximum ultimate bearing capacity of circular arch of SQCC150 × 8 is 1.05 MPa, which is 1.81 times that of U36.

(5) With the lateral pressure coefficient less than 0.5, the ultimate bearing capacity q_{1cr} of the circular arch is less than that of the straight-wall semi-circular arch, which is 0.3 to 0.6 times that of the latter. With the lateral pressure coefficient greater than 0.5, the former is significantly higher, about 1.5 to 15 times higher than the latter. It is even more obvious with the lateral pressure coefficient equal

to 1. In general, the ultimate bearing capacity of circular arches is higher than that of straight-wall semi-circular arches under the same conditions.

4.6 Chapter Summary

(1) The calculation and analysis model is established for the casing joints of the confined concrete arch. An analysis is made on the acting mode and the mechanical properties of the casing joints.

(2) The internal force calculation formula is deduced for the straight-wall semi-circular and circular arches with unequal rigidity and arbitrary sections. The influences are analyzed on the internal force of the arches by the factors such as the load q_1, the lateral pressure coefficient λ, the flexural rigidity of the arch EI, the equivalent joint rigidity ratio μ, the joint orientation angle α, and the height-diameter ratio κ.

(3) According to the calculating method for the ultimate bearing capacity of the confined concrete components, a comparison study is made on the ultimate bearing capacity of arches with different cross-sections. In general, the ultimate bearing capacity of circular arches is higher than that of straight-wall semi-circular arches under the same conditions.

Chapter 5
Experimental Study on the Bearing Mechanism of New High Strength Arches

In this chapter, a large-scale mechanical test system is designed and developed by the authors for the arch test. With this test system, a series of arch comparison tests are carried out; the deformation and the strain are obtained in the experiment process of the arches; and the arch bearing mechanism is clarified with different cross-section forms and shapes, and different load modes. The test results have verified the correctness of theoretical calculations and provides guidance for the design and the application of the new high strength arches.

5.1 Mechanical Test System for New High Strength Arches

5.1.1 Research and Development Background

In deep mines with the soft rock, failure of the traditional arches is common, and the support strength of roadways is facing the great challenge. With advantages of high strength and good economic performance, confined concrete arches are suitable for those roadways which are difficult to support, such as deep roadways with large deformation, rebuilt roadways after failure, expansive soft rock roadways, dynamic pressure roadways and crushed surrounding rock roadways. It is necessary to conduct mechanical tests on confined concrete arches of different shapes to obtain the ultimate bearing capacity of the arches and clarify their main mechanical properties. At present, there are a few similar model test benches, but they have the following shortcomings:

(1) The size of the model test benches is too small to perform full-scale or large-scale tests. The size of the supporting arch used in the field must be reduced to a certain scale to meet the requirement of the model test benches.
(2) They can perform tests on the arches only with a single shape and cannot meet the need for diversified arch shapes.

Q. Wang et al., *High Strength Support for Soft Surrounding Rock in Deep Underground Engineering*, https://doi.org/10.1007/978-981-15-3844-5_5

(3) With the loading operation fixed for one radius, they can only carry out a single-size arch test. So, they cannot be changed arbitrarily and their scope of use is limited.
(4) Most of their loading system cannot achieve the effect of flexible loading, so that the force transmitting head has a rigid contact with the surface of the arch. The stress concentration results in the arch failure, which affects the test accuracy.

In general, the underground projects have become increasingly complicated, especially in deep roadways, tunnels and chambers with the soft rock. The existing test bench systems can not meet the requirement. It cannot perform the tests of large sizes, multi shapes, high load, high hydraulic measurement and control accuracy, and accurate monitoring and acquisition. Therefore, a new large-scale mechanical test system is necessary for supporting arches in underground engineering to solve the above issues.

5.1.2 System Components and Main Functions

5.1.2.1 Research and Development Purposes

With the test system, relevant tests can be carried out to quantitatively obtain the ultimate bearing capacity and the main mechanical behavior of the supporting arches in underground engineering such as confined concrete arches. The system can reflect the true situation on site effectively, and analyze intuitively and deeply the deformation, the stability and the damage mechanism of the arches. It could provide the effective means for the mechanical test of large-scale chamber arches.

5.1.2.2 System Composition

This test system is mainly composed of the reaction force structure, the loading and controlling system, the monitoring system and auxiliary components, as shown in Fig. 5.1.

(1) The reaction force structure is the concrete structure with the large size (the outer diameter of the reaction force structure reaches 10 m), high strength and rigidity (can provide more than 2400*t* reaction force), and good stability. It can realize 1:1 mechanical tests on the arches of mine roadways or large-scale model tests of other large chamber arches; and it can also provide reaction forces for the test on the arches of different shapes through the assembly of modular modules.
(2) The loading and controlling system consists of a hydraulic pumping station, 12 groups of hydraulic cylinders, an automated measurement and control system, and a force transmission dispersing device.

Fig. 5.1 Test system

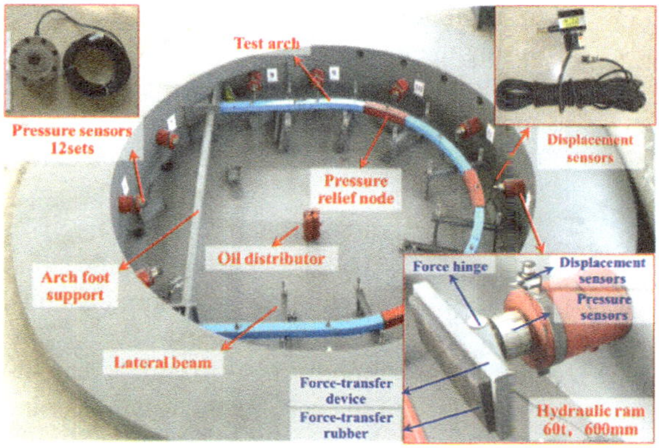

(1) The maximum pressure of the hydraulic pumping station is 20 MPa. The hydraulic pumping station and the controlling system can realize gradient loading—load holding—loading control on the 12 cylinders in two groups. The cylinders in each group are controlled proportionally to realize synchronous loading.
(2) The hydraulic cylinder is installed in a sliding groove, so the test can be carried out on confined concrete arches of different sizes.
(3) The automatic measurement and controlling system consists of data acquisition and processing, computer control, and the display system. It can realize high-speed sampling, test force and peak value in real time displaying. It also can set loading speed, hold targets and time in different stages, and output and draw various test curves as needed.
(4) The force transmission dispersing device is composed of a force transmission hinge, a force transmission disperser and a force transmission rubber. The force transmission hinge enables the force transmission disperser to ensure that the loading direction is perpendicular to the specimen contour when the specimen deformed. The force transmission disperser makes the cylinder thrust transmit to the larger extent on the arch and reduce stress concentration. The force transmission rubber makes the more full contact between the force transmission disperser and the arch surface, and therefore, it effectively prevents the component from early failure due to stress concentration, which affects the results of the experiment. The other accessory components referred mainly to the beams, which ensure that the tested arch deform in-plane when subjected to the test load.

(3) The monitoring system consists of monitors of radial force, angle, radial displacement, stress and deformation.

(1) Radial displacement monitoring means displacement sensors (the quantity is determined according to the need) are installed at the specified position of the specimen, and a data acquisition and processing unit is equipped

to accurately collect and analyze the radial deformation of the specimen during the test.

(2) Angle monitoring refers to provide the basic data for the specimen angle analysis at the specified location by arranging angle monitors at the designated position and by collecting and recording data through the acquisition unit.
(3) Radial force monitoring means that force sensors are installed on the loading head (12 in total) of each hydraulic cylinder with a data acquisition and processing unit equipped to accurately collect and analyze the radial force of the specimen during the test.
(4) Stress-strain monitoring is to collect data from the strain sensor attached to the surface of the specimen through a strain gauge, by which the effective analysis can be made on the micro-strain of a specified position of the specimen.

5.1.2.3 System Functions

(1) Realize the mechanical tests on confined concrete arches and other traditional arches.
(2) Combined with adjustment modules, this system can perform mechanical tests on confined concrete arches of different shapes; therefore, it is adapted to different shaped arches, such as round and straight wall semicircular; and it can be used under different working conditions of mine roadways, tunnels, hydropower tunnels and subways.
(3) By increasing or decreasing the number of blocks on the base of the cylinder, the effective loading radius of the test system can be adjusted to realize the test of arches with different sizes.
(4) Realize accurate measurement and collection of the test data on deformation, force, stress.

5.2 Experimental Study on the Bearing Mechanism of Circular Arches

In this section, a test on the typical circular arch is carried out to compare and analyze the bearing mechanism and mechanical properties of confined concrete and traditional supporting arches with different cross-sectional forms under the bias condition. It verifies the correctness of the previous theoretical analysis.

5.2.1 Experiment Scheme

Experiments are carried out to research the bearing mechanism of SQCC150 × 8, CCC159 × 10 and U36 arches with the same cross-sectional steel content under the bias condition of the lateral pressure coefficient $\lambda = 1.5$. The concrete core strength of SQCC and CCC arches is C40. The purpose of the research is to obtain the characteristics of stress, deformation, instability and failure of circular confined concrete arches under the biasing force; and it is also for comparing the mechanical properties of SQCC, CCC and U-steel arches, and analyzing the variation law of the load with the displacement and internal force distribution. Through the research, the correctness of the arch calculation theory is verified and the bearing properties of the arch is studied with different steel tube wall thickness, different concrete strength grades and different lateral pressure factors.

5.2.1.1 Experiment Processing

The inner diameter of the circular arches is 5.2 m. The three cross section forms are SQCC150 × 8, CCC159 × 10, and U36. Figure 5.2 shows the dimensions of SQCC150 × 8 and CCC159 × 10 arches. The specimen is made at the processing plant as required. Figure 5.3 shows the processed and assembled specimen of the SQCC arch. Figures 5.4 and 5.5 are the specimens of the processed and filled circular steel and U-steel arches respectively.

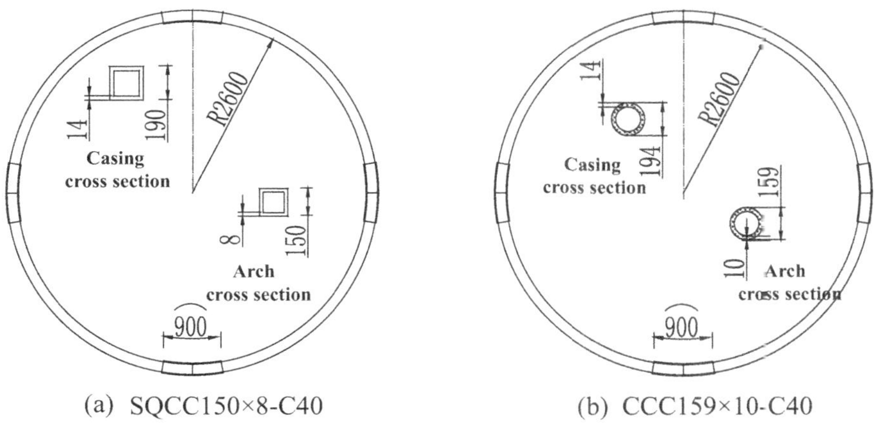

Fig. 5.2 Size of circular arch

Fig. 5.3 SQCC arch

Fig. 5.4 CCC arch

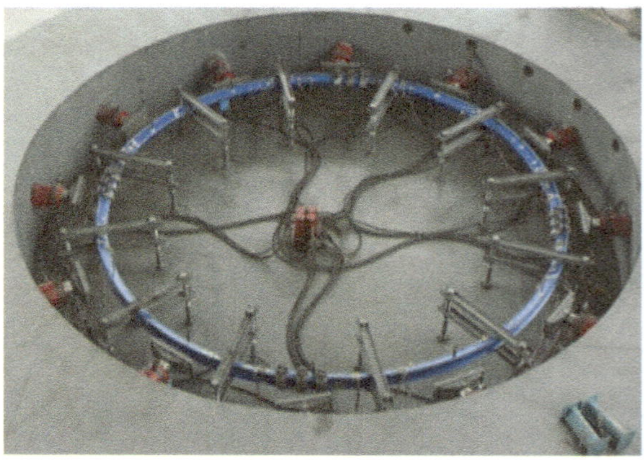

Fig. 5.5 U-steel arch

5.2.1.2 Loading and Monitoring Scheme

1. Loading Scheme

(1) Bias loading: three cylinders are on the top of the arch, they are 8#, 9# and 10#; three cylinders are at the bottom, they are 2#, 3# and 4#, three cylinders are on the left side, they are 5#, 6# and 7# and three cylinders are on the right side, they are 1#, 11# and 12#, as shown in Fig. 5.6. The load applied from the cylinders on the top and at the bottom of the arch is 1.5 times that from the cylinders at the two sides, that is, the lateral pressure coefficient $\lambda = 1.5$.

(2) Loading rate and holding time: Step loading method is adopted. When the load is less than 90% of the estimated ultimate load, the loading rate is 10 kN/min with the pressure holding time of 0.5 min per 30 kN; when the load exceeds 90% of the estimated ultimate load, the loading rate is reduced to 5 kN/min and the pressure holding time is 0.5 min per 10 kN; and the same setting is for bias loading.

(3) Stop loading standard: Keep loading with a monotonic pressurizing method until the arch failed. During the process, the damage of the arch is observed all time until the entire arch enters the yielding state or damaged.

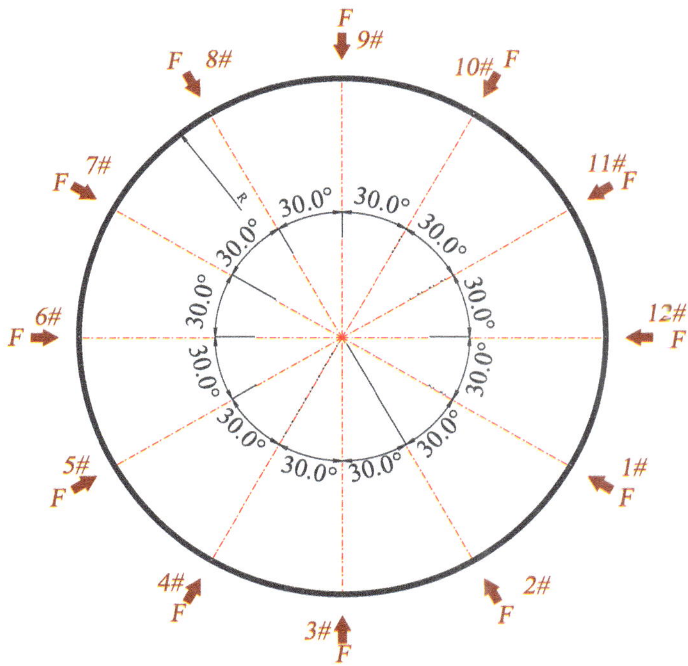

Fig. 5.6 Diagram of circular arch loading scheme

Table 5.1 Statistical table of monitoring items of circular arches

Monitoring content	Monitoring element	Quantity	Acquisition unit	Sampling frequency (s)	Number
Radial force monitoring	Spoke-Type force sensor 60*t*	12	Acquisition module	1	1#–12#
Radial displacement monitoring	Guyed displacement sensor 1000 mm	12	Acquisition module	1	1#–12#
Steel strain monitoring	Strain gauge 120-3CA	26 measuring points 78 slices	Static resistance strain gauge	2	Y1–Y26

2. Monitoring Scheme

In order to monitor effectively the force and the deformation of the arch and collect their information during the test, monitoring points are evenly arranged on the arch; and resistance strain gauges are arranged inside and outside the arch, and also on the sides of the arch at each monitoring point; and meantime, radial force monitoring and displacement monitoring are carried out at the loading point. The specific monitoring scheme is described in detail in each section. The monitoring information is shown in Table 5.1.

5.2.2 Numerical Test Scheme

Numerical tests are carried out to verify the conclusions of the experiment on the arches, and to complete the test data (such as arch axial force, bending moment, and concrete stress) which cannot be effectively collected in the experiment.

The tested arches of the three cross-sectional forms—SQCC150 × 8, CCC159 × 10 and U36 have the same cross-sectional steel content. The test size and the loading method is the same as the experiment.

1. Material Parameters

The material parameters in the numerical tests on the above-mentioned arch specimens are the same as those determined in the previous Sect. 3.1.1.2.

2. Boundary Conditions

Constraint is set in the tangential and the directions parallel to the axis of the arch at the sides of the arch. Tie constraint is used between the inner wall of the steel tube and the concrete core. Both the steel tube and the concrete core use three-dimensional solid elements, and the unit type is C3D8R or C3D20R, and the Molar Coulomb Failure Criterion is adopted.

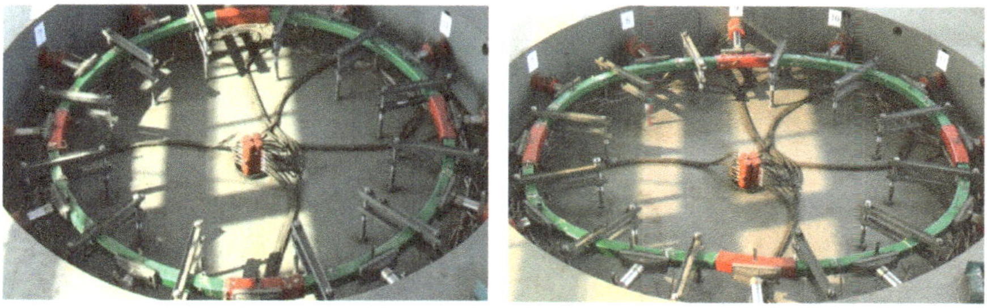

(a) Before experiment (b) After experiment

Fig. 5.7 SQCC arch deformation and failure modes

5.2.3 Arch Deformation and Failure Process

5.2.3.1 Deformation and Failure Process of the SQCC150 × 8-C40 Arch

Figure 5.7 is the morphology comparison of the SQCC arch before and after the experiment.

During loading, the following deformation and failure phenomena occur:

(1) At the beginning of the experiment, the deformation of the arch is small. As the load increases, the deformation becomes increasingly obvious. The faster increasing speed of the vertical load makes the overall shape of the arch changed from round to oval. The upper and lower sides are squeezed inward and the left and right sides are expanded outward.
(2) In about 631s, the phenomenon of coated layer bulging occurs on the arch near the loading point #3, indicating the first strength damage occurs at this point, and the bearing capacity of the arch has reached its limit at this moment.
(3) At the end of the experiment, serious damages occur on the vault, the bottom and at the two sides of the arch.

5.2.3.2 Deformation and Failure Process of the CCC159 × 10-C40 Arch

Figure 5.8 shows the overall and local deformation diagram of the CCC arch in the experiment.

During loading, the following deformation and failure phenomena occur on the arch:

(1) No obvious deformation of the arch is observed at the beginning of the experiment.
(2) As the load continues to increase, the faster increase speed of the vertical load makes the overall shape of the arch changed from round to oval. The upper

(a) The overall failure mode

(b) The local failure mode

Fig. 5.8 CCC arch deformation and failure modes

and lower sides are squeezed inward, and the left and right sides are expanded outward. The arch has been deforming along this law afterwards.

(3) The strength damage occurs near the bottom of the arch. With the occurrence of strength damages, the load starts to decrease, the deformation speed of the arch obviously accelerates, and damages begin to occur on the sides of the arch.
(4) By the end of the test, the deformation of the arch is more serious, and the most severe damages occur on the vault and the bottom, and at the two sides of the arch.

5.2.3.3 Deformation and Failure Process of the U36 Arch

1. Overall Deformation and Failure Process of the Arch
   Figure 5.9 shows the overall and local deformation failure mode of the U36 arch after the experiment.

   During loading, the following deformation and failure phenomena occur on the arch:

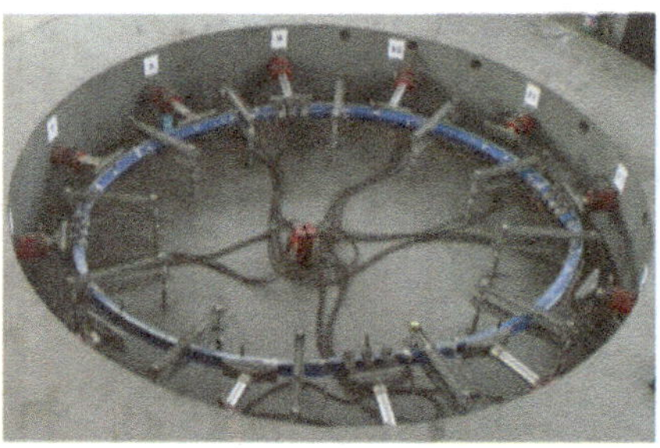

(a) The overall failure mode

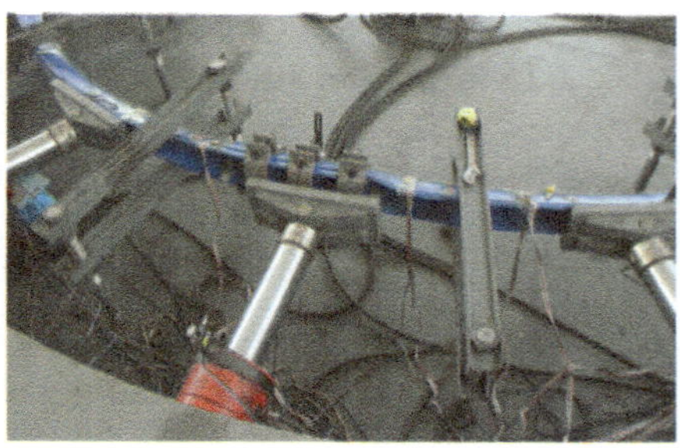

(b) The local failure mode

Fig. 5.9 U36 arch deformation and failure modes

(1) After the start of the experiment, the deformation of the arch increases uniformly with time, and then the deformation speed starts to increase, and the overall shape of the arch is changed from round to oval.
(2) In the late period of the experiment loading, the buckling failure occurs suddenly at the position of the cylinder #3, and the bearing capacity of the entire arch reduced drastically. More severe outward bending occurs at the position between the cylinders #12 and #1; inward bending occurs at the position of the cylinder #3; and more obvious strength damages occur at both of the positions.

2. Local Deformation and Failure Process of the Arch

(1) The local damage mainly located at the positions of #3 and #12. Especially, the out-plane buckling damage at the #3 position makes the integrity of the arch destroyed and the bearing capacity of the arch declined sharply.
(2) The arch has the buckling failure at the position #12. Although the out-plane buckling failure like #3 position does not occur here, the cross-section opening at this position expanded by 18 mm on both sides due to the bending force.

5.2.4 *Comparative Analysis on the Experiment Results*

5.2.4.1 Comparative Analysis on SQCC and CCC Arches

1. Comparison of Deformation and Failure Modes

(1) The first stage: At the beginning of the test, no obvious deformation is observed on the SQCC and CCC arches. As the load continues to increase, the overall shape of the arch is changed from round to oval. The upper and lower sides are squeezed inward, and the left and right sides are expanded outward.
(2) The second stage: The coated layer bulging phenomenon and damages occur on the SQCC arch near the loading point #3 at first, and the bearing capacity of the arch has reached its limit at this moment; and damages occur in the position near the bottom of the CCC arch, and the arch load has also reached its limit.
(3) The third stage: With the appearance of damages, the deformation speed of the arch obviously accelerated, and damages also start to appear at the sides of the arch.
(4) At the end of the experiment, the most severe damages occur on the vaults and the bottoms, and at the sides of the SQCC and CCC arches.

2. Analysis on the Arch Load Curve
Figures 5.10 and 5.11 show:

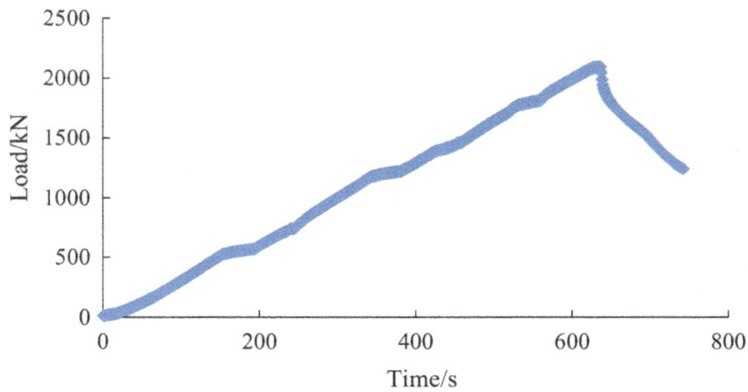

(a) Total load-time curve of the SQCC arch

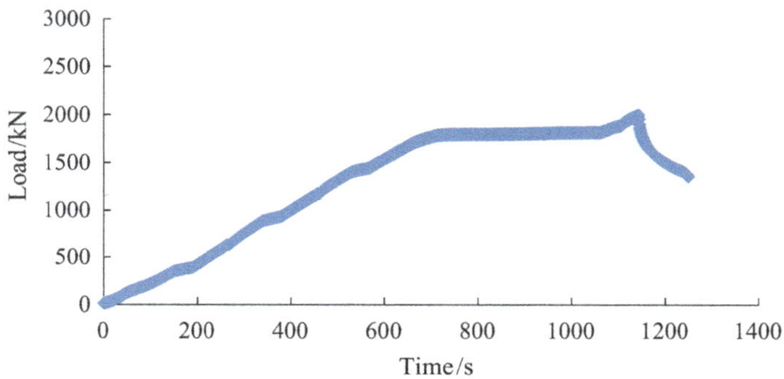

(b) Total load-time curve of the CCC arch

Fig. 5.10 Total load-time curve of the arches in the experiment

The ultimate bearing capacity of the SQCC arch is 2096.4 kN, the capacity of the CCC arch is 2003.2 kN, and the former is 1.05 times the latter. With the better bearing capacity, the SQCC arch can provide a higher radial force to the surrounding rock and maintain the stability of the surrounding rock, and therefore it is more conducive to the self-supporting ability of surrounding rock.

5.2.4.2 Comparative Analysis on SQCC and U-Steel Arches

1. Comparison on Deformation and Failure Modes

   (1) The first stage: the deformation of SQCC and U-steel arches increase uniformly with time, and then the deformation speed of both begin to accelerate. As the load continues to increase, the overall shape of the arches is changed from round to oval. The vault and bottom are squeezed inward, and the left and right sides are expanded outward.

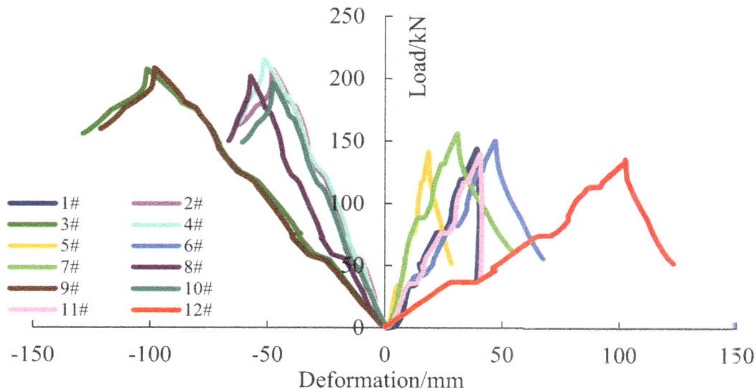

(a) Each cylinder load-displacement curve of the SQCC arch

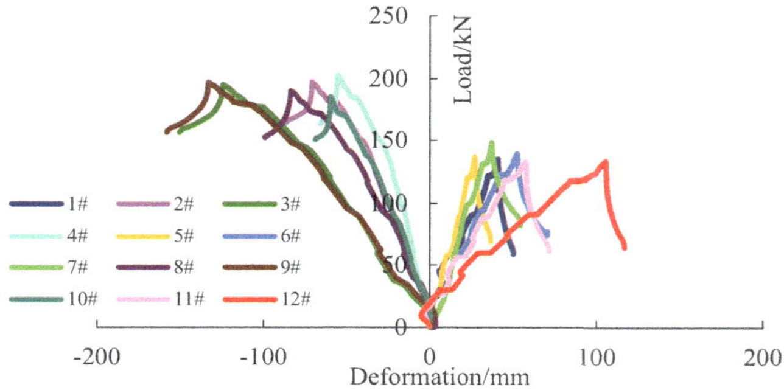

(b) Each cylinder load-displacement curve of the CCC arch

Fig. 5.11 Load-displacement curve of each cylinder

(2) The second stage: Until the late period of the test loading, a buckling failure occurs suddenly on the U-steel arch at the position of the cylinder #3, and the bearing capacity of the entire arch reduced drastically. More severe outward bending occurs at the position between the cylinders #12 and #1; inward bending occurs at the position of the cylinder #3; and more obvious strength damages occur at both of the positions. The phenomenon of coated layer bulging only occurs on the SQCC arch near the loading point 3#, indicating the first damage occurs at this position.

(3) The third stage: With the occurrence of strength damage, the bearing capacity of the arch begins to decrease, and the increasing deformation results in strength damages on the arch. With the faster deformation, the U-steel arch loses its bearing capacity, but the SQCC arch still have the high bearing capacity in the late stage.

(4) At the end of the experiment, the most severe strength damages occur on the vaults and the bottoms, and at the sides of both arches.

2. Analysis on the Arch Load Curve
   Figures 5.12 and 5.13 show:
   (1) The ultimate bearing capacity of the SQCC arch is 2096.4 kN; the capacity of the U- steel arch is 1198 kN; and the former is 1.75 times the latter. With the better bearing capacity, the SQCC arch can provide the higher radial force to surrounding rock and maintain the stability of the surrounding rock; and therefore it is more conducive to the self-supporting ability of the surrounding rock.
   (2) Under the influence of the bias load, the bearing capacity of both arches significantly reduced in the late stage. After the arches reach the ultimate bearing strength, their bearing capacity decreases rapidly as the experiment

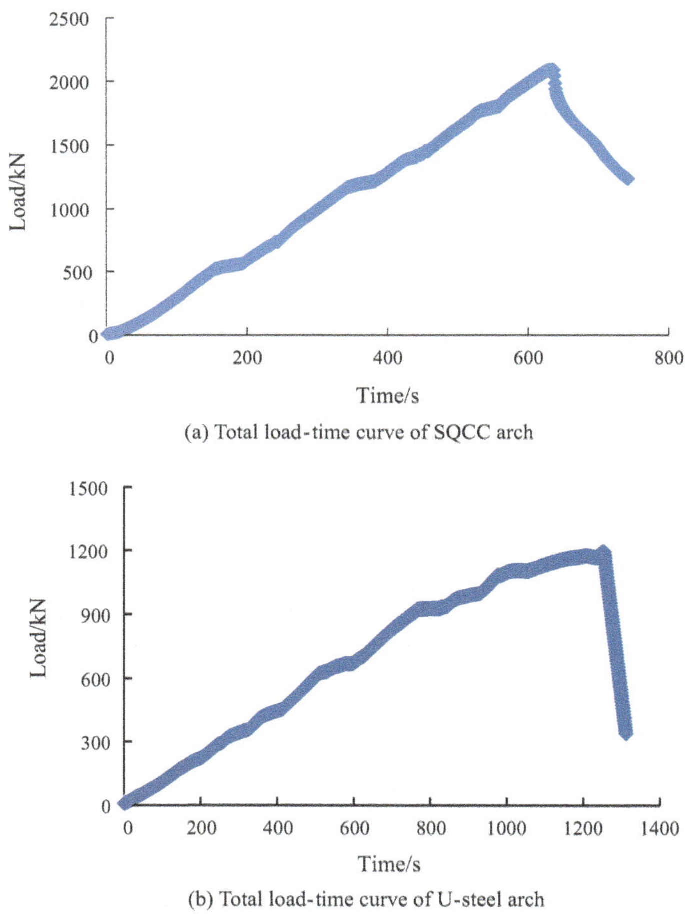

Fig. 5.12 Total load-time curves of the arches in the experiment

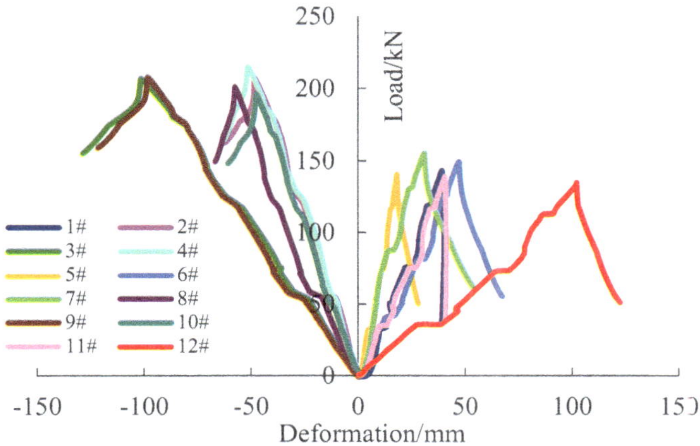

(a) Load-displacement curve of each cylinder of SQCC arch

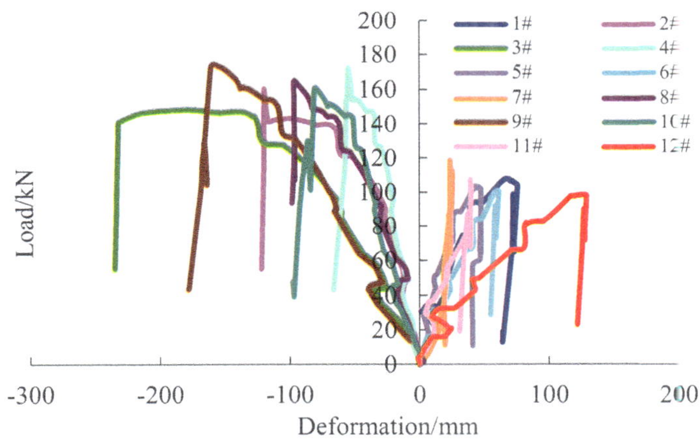

(b) Load-displacement curve of each cylinder of U-steel arch

Fig. 5.13 Load-displacement curve of each cylinder

continues. However, after the arch reaches its ultimate strength, the SQCC arch can still bear a certain load, but the U-steel arch completely loses its bearing capacity, indicating the ductility of the SQCC arch is better than that of the U-steel arch.

5.2.5 Analysis on the Internal Force of the SQCC Arch

According to the positions of the monitoring points shown in Fig. 5.6, the internal force of the cross-section of the arch is extracted in the corresponding positions from the numerical test of the SQCC arch; and the theoretical calculation is made on the internal force of the corresponding cross-sections and shown in Fig. 5.14.

Combined with the damage phenomenon of the arches in the experiment, the analysis is made through the internal force diagram of the arches:

(1) Define the axial force difference rate $\delta = \frac{|F_n - F_t|}{F_t} \times 100\%$ (F_n is the numerical test value, F_t is the theoretical calculation value). The maximum difference rate is 7.5% between the axial forces obtained by the theoretical calculation and the numerical test, and the minimum difference rate is only 1.5% between the two. Similarly, the maximum difference rate of the bending moment is 8.2% and the minimum difference rate is only 2.1% between the two. The internal force distribution of the two is quite consistent, which verifies the correctness of the theoretical calculation.
(2) The maximum axial force of the arch is 340.8 kN at the sides; the smallest axial force is 225.1 and 244.3 kN at the arch 0° (vault) and 180° (arch bottom), respectively. The conclusions of Chap. 3 show the ultimate load of the SQCC150 × 8 components is 2685 kN in the theoretical calculation, and the maximum axial force of the arch is only 12.7% of the ultimate load. The up and down distribution of the axial force has the good symmetry with the trend of increasing first and then decreasing from the vault to the side and then to the bottom of the arch.
(3) The bending moment on the vault and the bottom of the arch is negative, and bends the vault and bottom inward; and the bending moment on the sides is positive, and bends the sides outward. Therefore, during the experiment, the

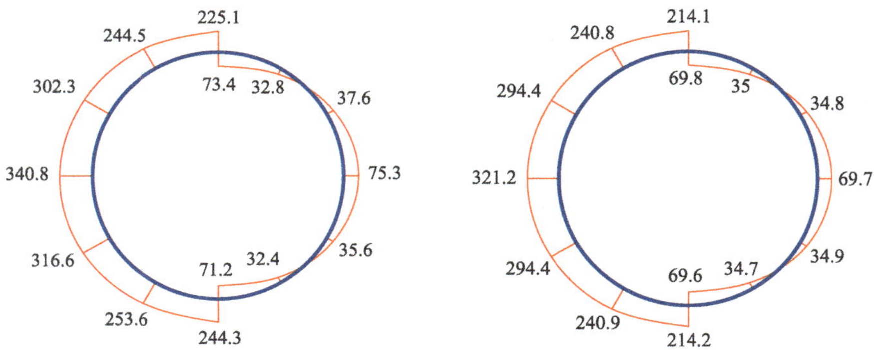

(a) Numerical test results (b) Theoretical calculation results

Fig. 5.14 Arch internal force distribution

overall shape of the arch changed from round to oval. The vault and bottom are squeezed inward, and the left and right sides are expanded outward.

(4) The bending moments are larger on the vault and the bottom, and at the two sides of the arch, therefore, serious damages occur in the four positions at the end of the test.

(5) The maximum bending moment at the sides is 75.3 kN m, and there are positions with the bending moment as 0 at the spandrel and the position near the bottom. The conclusions of Chap. 3 show the ultimate bending moment of the SQCC150 × 8 components is 100.2 kN m in the theoretical calculation, indicating the bending moments do not exceed their limit at the areas of the vault, the bottom and the sides of the arch. The analysis on the internal force of the arch shows the bending moment effect is more significant as the arch is damaged by the combined bending action under the bias loading.

5.2.6 *Analysis on Bearing Mechanism*

5.2.6.1 Influence Mechanism of Lateral Pressure Coefficients

Taking the SQCC arch as an example, a study is made on the different bearing mechanism of the confined concrete arch under different load modes by changing lateral pressure coefficients. The results show lateral pressure coefficients influence significantly on the ultimate bearing capacity of the SQCC150 × 8 arch, and the ultimate bearing capacity of the arch decreases with the increase of lateral pressure coefficients, as shown in Table 5.2.

The relation formula between the ultimate bearing capacity F_n of the arch and the lateral pressure coefficient λ is obtained by data fitting:

$$F_n = 6915.7\lambda^2 - 23,935\lambda + 22,568$$

where: $1.3 \leq \lambda \leq 1.7$, the fitting degree $R^2 = 0.996$.

Table 5.2 Bearing capacity statistics with different lateral pressure coefficient

Number	Lateral pressure coefficients	Bearing capacity (kN)	Reduction rate (%)
1	1.3	3162.3	0
2	1.4	2567.4	18.81
3	1.5	2234.1	29.35
4	1.6	2012.4	35.36
5	1.7	1845.8	41.63

Table 5.3 Bearing capacity statistics of arches with different concrete core strength

Number	Type of arch	Bearing capacity (kN)	Increase rate (%)
6	SQCC150 × 8-C30	2180.1	0
7	SQCC150 × 8-C40	2234.1	2.47
8	SQCC150 × 8-C50	2288.7	4.75
9	SQCC150 × 8-C60	2331.3	6.49
10	SQCC150 × 8-C70	2370.2	8.02

5.2.6.2 Influence Mechanism of the Concrete Core Strength

Taking the SQCC arch as an example, a study is made on the influence law of the concrete core strength on the bearing mechanism of the confined concrete arch by changing the concrete strength. The results are shown in Table 5.3.

The analysis shows the ultimate bearing capacity of the SQCC150 × 8 arch enhanced gradually with the improvement of the concrete core strength. The relation formula between the ultimate bearing capacity of the arch F_n and the concrete core strength $f_{\text{cu, k}}$ is obtained by data fitting:

$$F_n = -0.0336 f_{\text{cu,k}}^2 + 8.0831 f_{\text{cu,k}} + 1968.3$$

where: $30 \leq f_{\text{cu,k}} \leq 70$, the fitting degree $\text{R}^2 = 0.9999$.

5.2.6.3 Influence Mechanism of the Steel Tube Wall Thickness

The ultimate bearing capacity of the SQCC arch with different wall thickness under the lateral pressure coefficient $\lambda = 1.5$ is counted, as shown in Table 5.4.

The analysis shows that the ultimate bearing capacity of the SQCC150 × C40 arch increases linearly with the increase of the steel tube wall thickness. The relation formula between the ultimate bearing capacity of the arch F_n and the wall thickness t of the steel tube is obtained by data fitting:

$$F_n = 234.05t + 360.71$$

where: $7\text{ mm} \leq t \leq 11\text{ mm}$, the fitting degree $\text{R}^2 = 0.9954$.

Table 5.4 Bearing capacity statistics of arches with different wall thickness

Number	Type of arch	Ultimate bearing capacity (kN)	Reduction rate (%)
11	SQCC150 × 7-C40	2012.9	/
12	SQCC150 × 8-C40	2234.1	9.90
13	SQCC150 × 9-C40	2455.2	18.01
14	SQCC150 × 10-C40	2666.8	24.52
15	SQCC150 × 11-C40	2966.8	32.15

5.2.6.4 Analysis on the Ultimate Bearing Capacity

The numerical test is an improvement of the research on the bearing mechanism of the circular arch under the bias. The calculation results are shown in Table 5.5 for the ultimate bearing capacity of each calculation scheme of the circular arch.

(1) Under the bias, the ultimate bearing capacity of the SQCC150 × 8-C40 arch is 74.9% higher than that of the U36 arch with the same cross-sectional steel content. Compared with the steel arch, the bearing capacity of the SQCC arch is much higher.
(2) Under the bias, the ultimate bearing capacity of the SQCC150 × 8-C40 arch is 11.1% higher than that of the CCC159 × 10-C40 arch; the SQCC arch has a better associativity with the concrete spray layer; the sections of the arch can

Table 5.5 The bearing capacity of circular arch

Test sequence number	Arch model	Load type	Ultimate bearing capacity		
			Experiment (kN)	Numerical test (kN)	Difference rate (%)
1	SQCC150 × 8-C40	Lateral pressure coefficient λ = 1.5	2096.4	2234.1	5.6
2	CCC159 × 10-C40	Lateral pressure coefficient λ = 1.5	2003.2	1820.7	9.1
3	U36	Lateral pressure coefficient λ = 1.5	1198.4	1254.2	4.7

be connected easily; and the SQCC arch is more stable with better bending performance. It is more applicable in underground engineering.

5.3 Experimental Study on the Bearing Mechanism of Straight-Wall Semi-circular Arches

In this section, the test on typical straight-wall semi-circular arches is carried out to compare and analyze the bearing mechanism and mechanical properties of confined concrete arches and traditional arches with different cross-section forms under uniform pressure. The study has verified the correctness of the previous theoretical analysis.

5.3.1 Experiment Scheme

A comparative study is made on mechanical properties of SQCC, CCC and U-steel arches to obtain the stress, deformation, instability and failure behavior of the straight-wall semi-circular confined concrete arch under uniform pressure. In the study, the variation laws of load, displacement and internal force distribution of the arches are analyzed, and the correctness of the arch calculation theory is verified; and combined with the numerical tests, the bearing mechanism of arches is studied under different steel tube wall thickness, different concrete strength grades and different lateral pressure factors. The bearing mechanism of SQCC150 × 8, CCC159 × 10 and U36 arches is also studied with the same cross-sectional steel content and the uniform pressure loading. The concrete core strength of SQCC and CCC arches is C40.

5.3.1.1 Experiment Processing

The arch size in the experiment is the same as that in the field test scheme. The U-steel arch has the net width of 5500 mm, a straight wall of 1650 mm high, the joint overlap length of 500 mm and three pairs of high-strength clamps used for connecting. The SQCC and CCC arches have the net width of 5000 mm and a straight wall of 2000 mm; and the joints are connected by casings. The parameters of arches in the test are shown in Figs. 5.15 and 5.16.

The specimen is processed and formed at the processing plant as required at first. Then the concrete core should be poured in time, an attention should be paid to vibrating and compacting during pouring, and the specimen should be well maintained after pouring. Figure 5.17 shows the test arch.

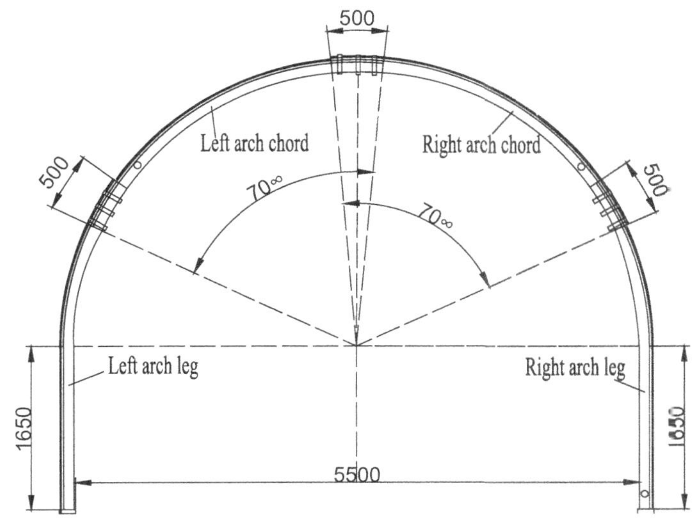

Fig. 5.15 Size of U-steel arch

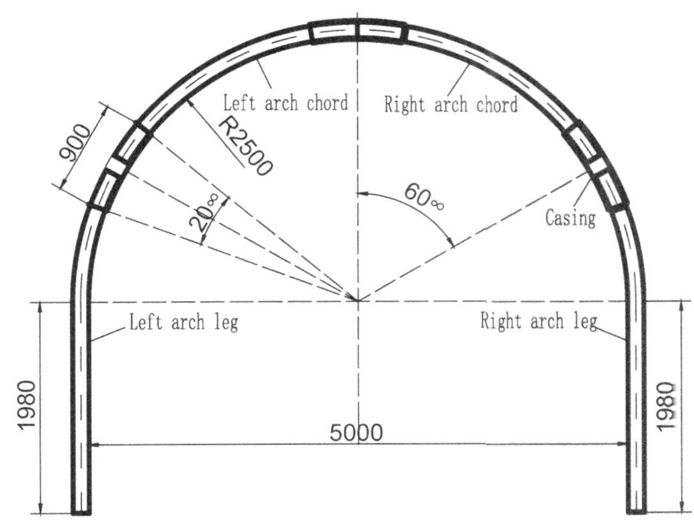

Fig. 5.16 Size of SQCC and CCC arch

5.3.1.2 Loading and Monitoring Scheme

1. Loading Scheme

   (1) Loading rate and holding time: Step loading is used. When the load is less than 90% of the estimated ultimate load, the loading rate is 10 kN/min with

Fig. 5.17 Test arch

the holding time of 0.5 min per 30 kN; when the load is greater than 90% of the estimated ultimate load, the loading rate is 5 kN/min with the holding time of 0.5 min per 10 kN; and the same is for the bias loading.

(2) Stop loading standard: Keep loading by monotonic pressurization until the specimen is destroyed. In the entire process, the damages of the specimen are observed at all times until the entire specimen enters the yielding state or is significant damaged.

2. Monitoring Scheme
   In order to effectively monitor and collect the force and deformation data in the arch test, the monitoring points are arranged according to the positions shown in Fig. 5.18. The specific monitoring information is shown in Table 5.6.

5.3.2 Numerical Test Scheme

Numerical tests are carried out on SQCC150 × 8, CCC159 × 10 and U36 arches with the same size and loading method in the experiment. All the arches have the same cross-sectional steel content.

1. Material Parameters
   In the above numerical tests, the material parameters are the same as those determined in the foregoing.
2. Boundary Conditions
   Constraint is set at the side of the arch in the direction parallel to the axis of the arch and also at the medial side of the skewbacks in the other directions of the plane. Tie constraint is used between the inner wall of the steel tube and the concrete core. Three-dimensional solid elements are adopted for both the steel tube and the concrete core with the unit type as C3D8R or C3D20R. The Molar Coulomb Failure Criterion is adopted.

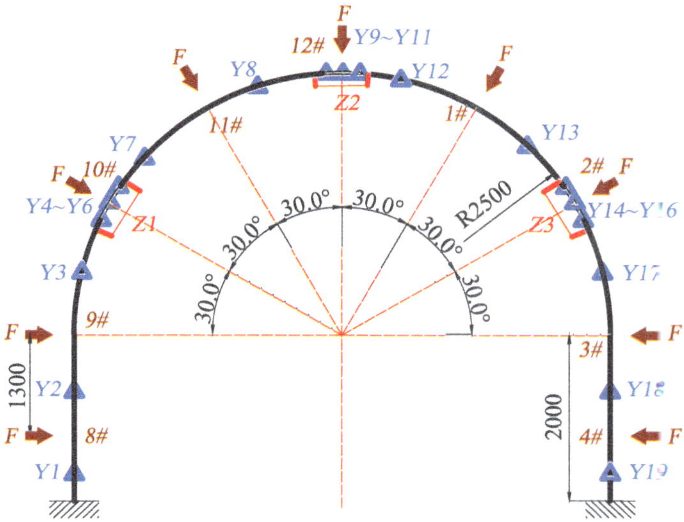

(a) SQCC and CCC arches

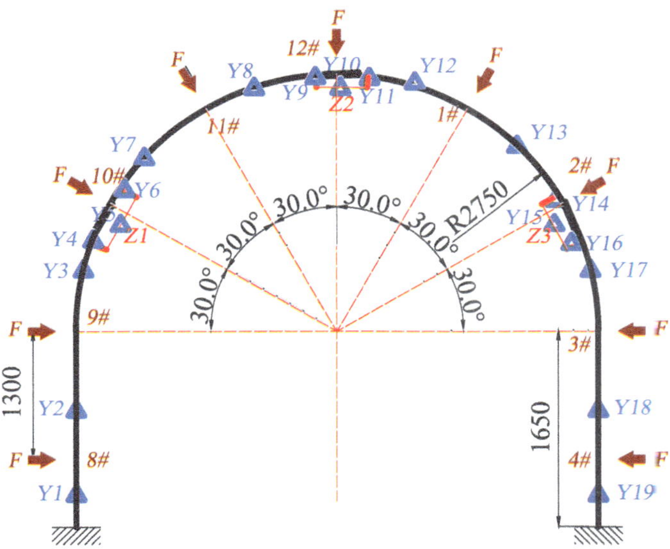

(b) U36 arch

Fig. 5.18 Schematic diagram of monitoring points

Table 5.6 Monitoring program

Monitoring content	Sensor	Quantity	Acquisition unit	Sampling frequency (s)	Number
Radial force monitor	Spoke-type force sensor 60*t*	9	Acquisition module	1	1#–12#
Radial displacement monitor	Guyed displacement sensor 1000 mm	9	Acquisition module	1	1#–12#
Steel strain monitor	Strain gauge 120-3CA	13 measuring points/33 slices	Static resistance strain gauge	2	Y1–Y19

5.3.3 Arch Deformation and Failure Process

5.3.3.1 Deformation and Failure Process of the SQCC150 × 8-C40 Arch

Figure 5.19 is the morphology comparison of the SQCC arch before and after the experiment.

During loading, the following deformation and failure phenomena occur on the arch:

(1) No obvious deformation is observed on the arch for a long period of time after the experiment starts.
(2) As the load continue to increases, the overall shape of the arch becomes thinner, the left and right arch legs are squeezed inward, and the arch vault is convex outward.
(3) As the load increases, the bending deformation becomes more obvious at the arch legs, which continues to squash inward and the vault is further convex outward. At this time, the arch has entered the yielding state.

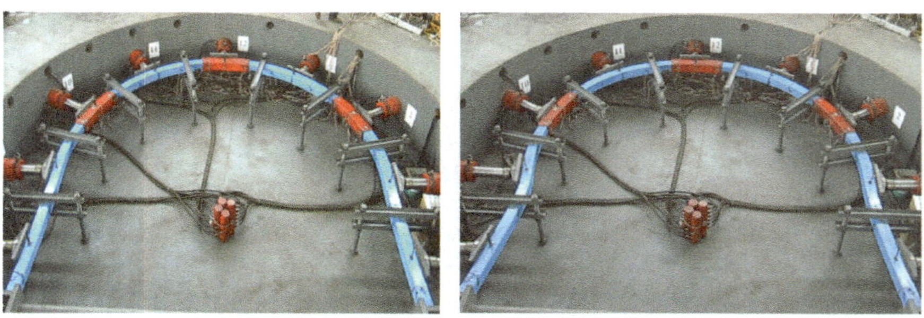

(a) Before experiment (b) After experiment

Fig. 5.19 SQCC arch deformation and failure modes

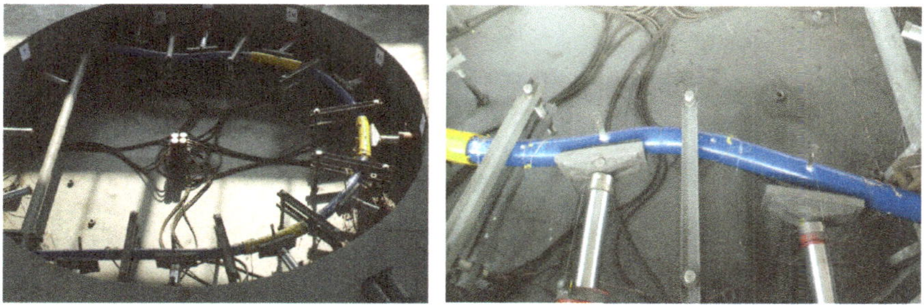

(a) Overall failure mode (b) Left arch leg failure mode

Fig. 5.20 CCC arch deformation and failure modes

(4) When the cylinder load reaches its maximum, the bending deformation at the arch legs is obvious, and the left is more significant.
(5) By the end of the experiment, the deformation of the arch is more serious and the maximum deformation occurs at the left arch leg.

5.3.3.2 Deformation and Failure Process of the CCC159 × 10-C40 Arch

Figure 5.20 shows the overall and local deformation of the CCC arch in the experiment.

During loading, the following deformation and failure phenomena occur on the arch:

(1) No obvious deformation of the arch is observed in the initial stage of the experiment.
(2) As the load continues to increase, the overall shape of the arch becomes thinner, the left and right arch legs are squeezed inward, the vault is convex outward, and the arch enters the yielding state.
(3) Then, the bending deformation at the arch leg is obvious, and the left is more significant; and at the end of the experiment, the deformation of the arch is more serious, and the maximum deformation occurs at the left arch leg.

5.3.3.3 Deformation and Failure Process of the U36 Arch

1. Overall Deformation and Failure Process of the Arch

   Figure 5.21 shows the overall and local deformation and failure modes of the U36 arch after the experiment.

   During loading, the following deformation and failure phenomena occur on the arch:

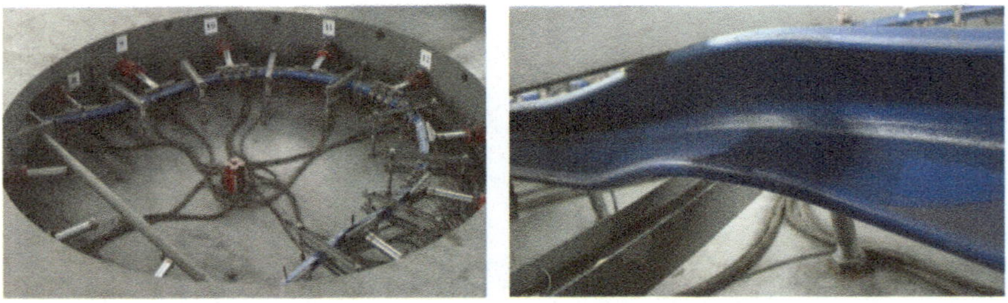

(a) Overall failure mode (b) Local failure mode

Fig. 5.21 U36 arch deformation and failure modes

(1) Under the uniform load, the arch deformed slowly for a long period of time after the experiment starts, and the cylinder pushes evenly.
(2) In about 782s, the overall shape of the arch becomes thinner, the left and right arch legs are squeezed inward, the vault is bent downward, and the arch has entered the yielding state.
(3) In about 820s, the right arch leg buckled suddenly and as a result, the entire arch loses its stability and the load has a sudden sharp drop. By the end of the experiment, the deformation of the arch becomes more serious, and the bending deformation is more obvious at the arch legs and on the vault.
(4) In about 1115s, the experiment is over and the maximum deformation occurs at the arch legs.

2. Local Deformation and Failure Process of the Arch

(1) After the test, serious "Z"-type flexural-torsional failure occurs at the left and right arch legs; the "U"-shaped cross-section is severely damaged; and the entire arch becomes unstabilized out-plane and losts its bearing capacity totally. The left and right deformation are basically symmetrical.
(2) At the same time, an obvious peeling of the coated layer occurs at the arch legs.

5.3.4 Comparative Analysis on the Experiment Results

5.3.4.1 Comparative Analysis on SQCC and CCC Arches

1. Comparison of Deformation and Failure Modes

(1) The first stage: no obvious deformation of the arch is observed at the beginning of the experiment. As the load continues to increase, the left and right arch legs are squeezed inward. About 10 min after the experiment starts,

the bending deformation is more obvious on the CCC arch legs than on the SQCC arch legs.

(2) The second stage: in about 15 min, the arch legs continue to squeeze, and the SQCC and CCC arches enter the yielding state.

(3) The third stage: In about 30 min, the cylinder reaches the maximum loading value; the SQCC and the CCC arches reach their maximum deformation value; and the largest deformation occurs at the arch legs.

2. Analysis on the Arch Load Curve

Figure 5.22 shows the variation curve of the total load over time for the nine groups of the cylinders obtained from the SQCC and CCC arch experiments. Figure 5.23 shows the radial load-displacement curves for each cylinder in the SQCC and CCC arch tests.

As Figs. 5.22 and 5.23 show:

(1) The ultimate bearing capacity of the SQCC arch is 1286.9 kN, the ultimate bearing capacity of the CCC arch is 1072.4 kN, and the former is 1.2 times

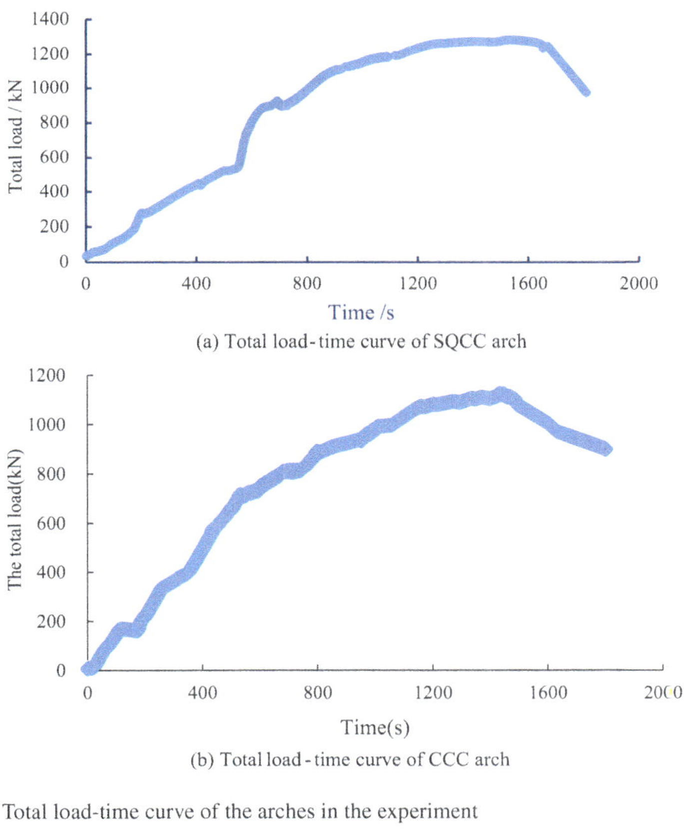

Fig. 5.22 Total load-time curve of the arches in the experiment

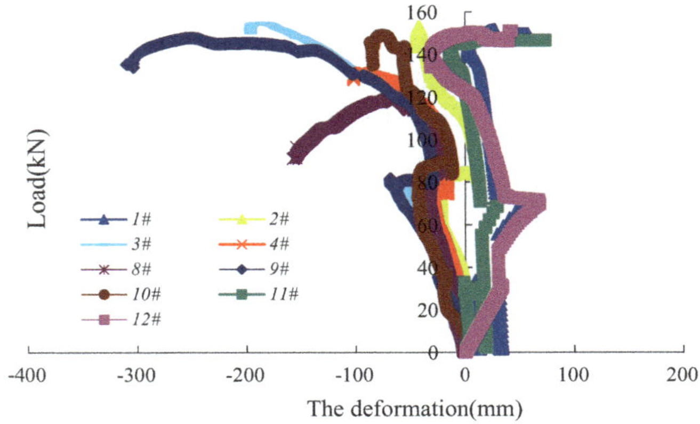

(a) Load-displacement curves of each cylinder of the SQCC arch

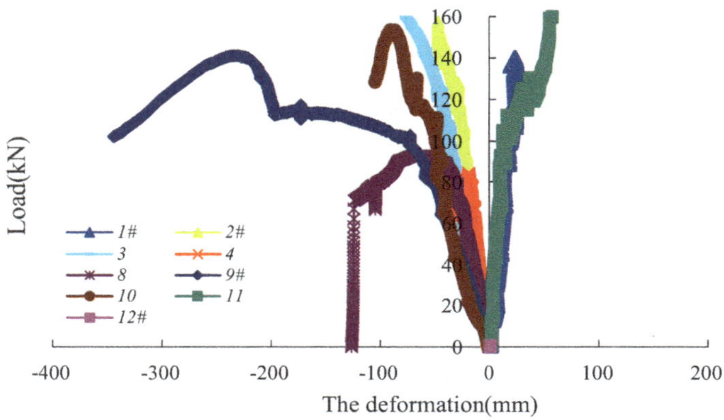

(b) Load-displacement curves of each cylinder of the CCC arch

Fig. 5.23 Load-displacement curves for each cylinder

the latter. The SQCC arch have the much better bearing capacity than the CCC arch. Meanwhile, the SQCC arch can provide the higher radial force to surrounding rock and effectively maintain the stability of surrounding rock, and so, the surrounding rock could better play its self-supporting ability.

(2) After the SQCC arch reaches the ultimate bearing strength, as the experiment continues, the load arch born decreases slowly, and the bearing capacity does not drop suddenly. Even if the arch deformed greatly, it still has the high load bearing capacity. The declining rate of the bearing capacity of the SQCC arch is much lower than that of the CCC arch, indicating the SQCC arch has the better bearing capacity in the late stage.

5.3.4.2 Comparative Analysis on SQCC and U-Steel Arches

1. Comparison of Deformation and Failure Modes
   The experiment on SQCC and U-steel arches can be divided into four stages:

   (1) The first stage: under the uniform load, the arches deformed slowly for a long period of time after the experiment starts, and the cylinders push evenly.
   (2) The second stage: the overall shape of the arches becomes thinner, the left and right arch legs are squeezed inward, the SQCC arch vault is convex outward and the U-steel arch vault is bent downward, and the both arche enters the yielding state. The bending deformation is significant on the arch legs of the U-steel arch, but the deformation is small on the SQCC arch.
   (3) The third stage: the right leg of the U-steel arch buckled suddenly, and as a result, the entire arch loses its stability and the load drops suddenly and sharply. By the end of the experiment, the deformation of the arch is more serious, and the bending deformation is more obvious on the vault and at the arch legs. In contrast, the SQCC arch deformed uniformly with arch legs squeezed inward and its vault convex outward until the arch enters the yielding state.
   (4) The fourth stage: at the end of the experiment, the maximum deformation occurs at the arch legs.

2. Analysis on the Arch Load Curve
   Figure 5.24 shows the variation curve of the total load-time for the nine groups of the cylinders in the experiment of the SQCC and U-steel arches. Figure 5.25 shows the radial load-displacement curve for each cylinder in the experiments of the SQCC and U-steel arches.

As Figs. 5.24 and 5.25 show:

(1) The ultimate bearing capacity of the SQCC arch is 1286.9 kN, the ultimate bearing capacity of the U-steel arch is 850.9 kN, and the former is 1.5 times the latter. The SQCC arch has the much better bearing capacity than the U-steel arch, so it can provide a higher radial force to the surrounding rock, maintain effectively the stability of the surrounding rock, and therefore, the surrounding rock could better play its self-supporting ability.
(2) After the arch reaches the ultimate bearing strength, as the test continues, the declining rate of the bearing capacity of the SQCC arch is much lower than that of the U-steel arch, indicating the SQCC arch has the better bearing capacity in the late stage.

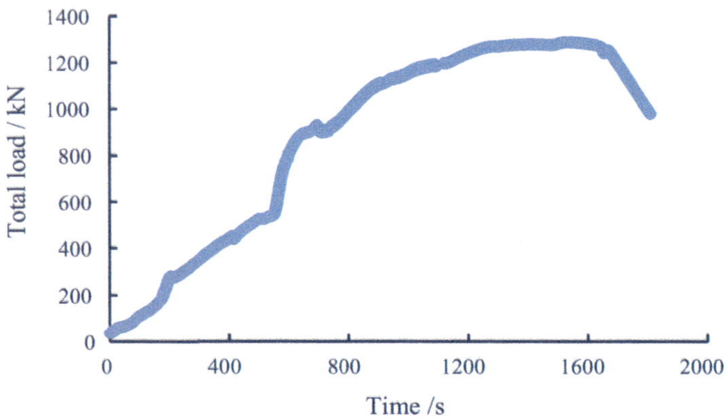

(a) Total load-time curve of the SQCC arch

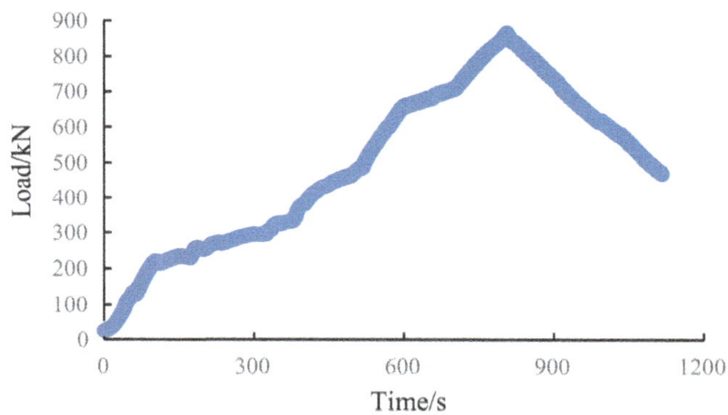

(b) Total load-time curve of the U-steel arch

Fig. 5.24 Total load-time curves of the arches in the experiments

5.3.5 Analysis on the Experiment Results of the SQCC Arch

When the ultimate bearing capacity reached, the axial force and the bending moment of the typical cross-section of SQCC arches are extracted from the numerical test and the theoretical calculation, and plot the values in Fig. 5.26.

The analysis shows:

(1) The axial force and the bending moment diagrams of the arch in the numerical test are basically the same as those in the theoretical calculation of the fourth chapter in shape which verifies the correctness of the theoretical calculation. The largest bending moment is 116.9 kN m at the arch legs as the ultimate bearing capacity reached. This position is in the worst state.

(2) Under the uniform load, the axial force of the SQCC arch increased gradually from the skewback to the vault, but in general, the difference is little, and the

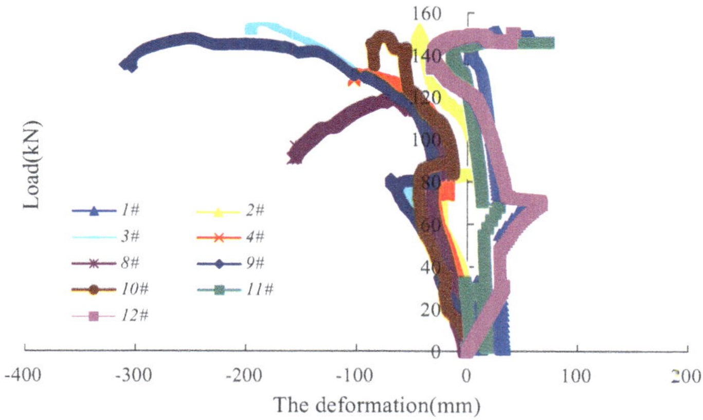

(a) Load-displacement curve of each cylinder of the SQCC arch

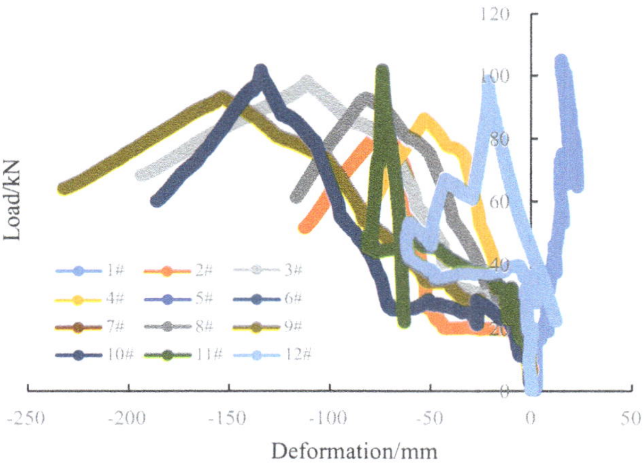

(b) Load-displacement curve of each cylinder of the U-steel arch

Fig. 5.25 Load-displacement curves for each cylinder

largest axial force located on the vault. This value is only 12% of the ultimate bearing capacity of the SQCC short column components under axial compression. The position of maximum bending moment is 1–2 m from the arch skewback; the direction of the bending moment is positive; and the positive bending moment makes the arch legs squeezed inward. All of these are consistent with the failure patterns in the experiment, indicating the arch failure is caused mainly by bending moment.

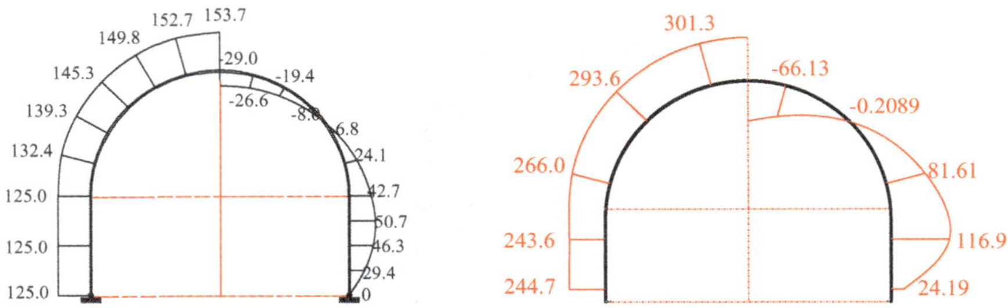

(a) Theoretical calculation (b) Numerical test

Fig. 5.26 Arch internal force distribution

5.3.6 Analysis on Bearing Mechanism

5.3.6.1 Influence Mechanism of Lateral Pressure Coefficients

Taking the SQCC arch as an example, the bearing mechanism of the confined concrete arch is studied under different load modes by changing lateral pressure coefficients.

The results show: as the lateral pressure coefficient $\lambda = 0.5$, 1.0 and 2.0, the ultimate bearing capacity of the arch decreases obviously with the increase of the lateral pressure coefficient; as λ is 1.0 and 2.0, the ultimate bearing capacity of the SQCC150 × 8 arch is 59 and 39% of that as λ is 0.5. Therefore, it is necessary to analyze carefully the on-site stress conditions to select and design the arch, and try to avoid the straight wall semi-circular arch under large lateral pressure.

5.3.6.2 Influence Mechanism of the Concrete Core Strength

Firstly, the influence of the concrete core strength is studied. Different types of concrete are filled in the SQCC150 × 8 arches and the corresponding ultimate bearing capacity is obtained, as shown in Table 5.7.

From the above results, the following analysis is made:

(1) With the increase of the concrete strength, the ultimate bearing capacity of the SQCC arch increases. However, this influence is not obvious, and the ultimate bearing capacity of the SQCC150 × 8-C80 arch is increased by 7.89% comparing with that of SQCC150 × 8-C30.
(2) By data fitting, the relationship between the ultimate bearing capacity of the SQCC arch (F_n) and the concrete core strength ($f_{\mathrm{cu.k}}$) is obtained and expressed as:

Table 5.7 Bearing capacity statistics of arches with different concrete core strength

Number	Type of arch	Ultimate bearing capacity (kN)	Increase rate (%)
16	SQCC150 × 8-C30	1217	/
17	SQCC150 × 8-C40	1248	2.55
18	SQCC150 × 8-C50	1275	4.77
19	SQCC150 × 8-C60	1303	7.07
20	SQCC150 × 8-C70	1309	7.56
21	SQCC150 × 8-C80	1313	7.89

$$F_{\rm n} = \begin{cases} 2.8389 f_{\rm cu,k} + 1113 \ \ (30 \le f_{\rm cu,k} < 60) \\ -0.0081 f_{\rm cu,k}^2 + 1.6343 f_{\rm cu,k} + 1234.1 \ \ (60 \le f_{\rm cu,k} < 80) \end{cases}$$

When $30 \le f_{cu,k} < 60$, the fitting goodness is 0.9972, and when $60 \le f_{cu,k} \le 80$, the fitting goodness could be up to 1.

5.3.6.3 Influence Mechanism of the Steel Tube Wall Thickness

Next, the influence of the steel tube thickness is analyzed on the ultimate bearing capacity. In the numerical simulation, different steel tube thicknesses are adopted and the corresponding ultimate bearing capacity is obtained, as shown in Table 5.8.

Table 5.8 The bearing capacity of arches with different thickness

Number	Type of arch	Ultimate bearing capacity (kN)	Reduction rate (%)
22	SQCC150 × 4-C40	776	/
23	SQCC150 × 6-C40	1007	29.73
24	SQCC150 × 8-C40	1248	60.72
25	SQCC150 × 10-C40	1466	88.81
26	SQCC150 × 12-C40	1600	106.11

Table 5.9 Statistical table of arch bearing capacity

Test sequence number	Type of arch	Load type	Ultimate bearing capacity		
			Experiment (kN)	Numerical test (kN)	Difference rate (%)
1	SQCC150 × 8-C40	Uniform pressure	1286.9	1248	3.02
2	CCC159 × 10-C40	Uniform pressure	1072.4	1068	0.41
3	U36	Uniform pressure	/	794	/

The table shows the ultimate bearing capacity of the SQCC arch increases as the steel tube thickness increases. Their relationship could be expressed as:

$$F_n = -0.9768t^3 + 19.597t^2 - 7.4075t + 555.49$$

where $4 \le t \le 12$, F_n represents the ultimate bearing capacity, and the fitting goodness could be up to 1.

5.3.6.4 Analysis on the Ultimate Bearing Capacity

The bearing mechanism and mechanical properties of the confined concrete arch are further clarified by the numerical test. The calculation results of the ultimate bearing capacity of the straight wall semi-circular arches in the calculation schemes are shown in Table 5.9.

The results show the ultimate bearing capacity of the SQCC150 × 8 and CCC159 × 10 arches are 1.3–1.6 times that of the U36 arch. It is a new high strength support form for soft surrounding rock in deep underground.

5.4 Project Suggestions

(1) The ultimate bearing capacity of the SQCC arch is 1.05–1.2 times that of the CCC arch and 1.5–1.75 times that of the U-steel arch. The SQCC arch has a higher radial reacting force on the surrounding rock and it is more conducive to the self-supporting ability of the surrounding rock.
(2) The failure modes of SQCC and CCC arches are basically the same. No obvious strength damage occur under larger deformation. They have the better post bearing capacity and the effective control capability on the surrounding rock.
(3) The SQCC arch has the larger inertia moment than the CCC arch does. It has the better flexural properties, the larger contact area with the surrounding rock and the higher bearing capacity. In addition, it is more evenly stressed and it is

easy to apply longitudinal connection and form the supporting system during field construction. Therefore, the SQCC arch is recommended in underground engineering.

(4) In the field application, to reduce the influence of bending moment, tie rods should be installed between the arches for the longitudinal connection or pre-stressed steel strands should be arranged between the arch and the anchor (cable) to increase the force fulcrums and reduce the bending moment on the arch.
(5) In the field application, it is recommended to optimize the cross-section shape of the chamber, weld the retaining plate at the legs (the key failure positions) of the arch, reduce the length of the arch leg and increase the quantities of longitudinal connections.

5.5 Chapter Summary

(1) A large-scale mechanical test system is designed and developed for the new high strength arches in underground engineering. A series of full-scale comparative experiments are conducted on the confined concrete arches of U36, SQCC150 $\times$ 8 and CCC159 $\times$ 10. The influence factors and laws are studies on the mechanical properties of the confined concrete arches.
(2) The bearing mechanism of the SQCC arch is analyzed, and the influence mechanism of the concrete core strength, the lateral pressure coefficient and the steel tube wall thickness is discussed on the ultimate bearing capacity of arches.

Chapter 6
Engineering Application of the New High Strength Support in Soft Rock Roadways in the Sea Area

Based on design and construction methods proposed in this book, the new high strength support system is applied in the underground engineering under complex conditions such as high stress and soft rock. This chapter takes the Liangjia Coal Mine, the extremely soft rock roadway in the sea area, as an engineering example. The failure of the bearing structure and the surrounding rock control mechanism of typical deep soft rock roadways are clarified through the field monitoring and numerical tests. Therefore, the confined concrete support system is verified to be rational and effective in surrounding rock control. This research could provide guidance for the design and field application of deep soft surrounding rock support.

6.1 Engineering Background

Liangjia Coal Mine is adjacent to China Bohai Sea and in Longkou Mining Area of Shandong Province, China, as shown in Fig. 6.1. It is a typical deep mine with soft rock and also the largest coastal coal mine in China. It has the designed production capacity of 2.8 million tons per year and its mining depth is −80 ~ −960 m.

The stratum structure of Liangjia Coal Mine is extremely complex and unstable. The minefield is full of fault structures and the normal faults are relatively developed. Among the faults, 70 faults have the drop greater than or equal to 20 m, 49 faults have the drop of 10–20 m, 50 faults have the drop of 5–10 m, and 56 faults have the drop of 3–5 m. Generally, the fault dip angles are 50°–70° and the directions of the faults are mainly east to west, northeast to east and northeast to northwest. The main coal-bearing strata are the Paleogene Lijiaya Formation in the mine. The lithology of the seam roof and floor rock is mainly mudstone, oil-bearing mudstone, sandy mudstone and oil shale. The surrounding rock belongs to medium-strong expansive soft rock according to the expansive soft rock classification standard. It is very difficult to support.

Q. Wang et al., *High Strength Support for Soft Surrounding Rock in Deep Underground Engineering*, https://doi.org/10.1007/978-981-15-3844-5_6

Fig. 6.1 Location of Liangjia Coal Mine

With a circular cross section, the roadway is originally supported with bolt-shotcrete + U36 arch, as shown in Fig. 6.2. The bolt type is Φ25 × 2250 mm with inter-row spacing of 650 × 800 mm; C20 concrete is sprayed on the roadway surface with the thickness of 100 mm; and the arches are made of U36 steel with the row spacing of 800 mm.

Due to the difficulty of surrounding rock control, several destruction phenomena occur even under such the strong combined support, such as floor heave, severe roadway contraction and support component failure, as shown in Fig. 6.3.

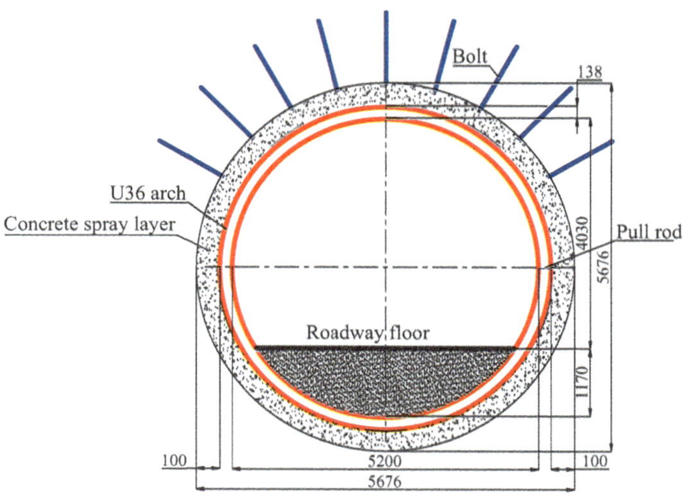

Fig. 6.2 Original supporting program (mm)

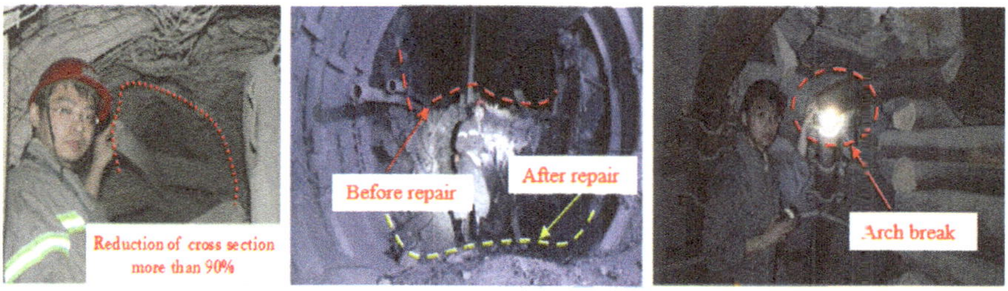

Fig. 6.3 Roadway destruction

6.2 Failure Mechanism of the Bearing Structure in Soft Rock Roadways

6.2.1 Deformation and Failure Mechanism of Surrounding Rock

1. Monitoring on Convergence of Surrounding Rock

The roof, floor and two sides of the roadway are monitored emphatically. The results are shown in Fig. 6.4.

The monitoring result shows the surrounding rock has the large and continuous deformation with poor stability around the roadway under the original supporting scheme. Deformation occurs immediately after the completion of the roadway excavation. The convergence of the two sides has reached to 249 mm in total in the 9th day of monitoring; and the convergence has reached to 527 mm with deformation of 397 mm on the roof and floor of the roadway in the 20th day. Throughout the monitoring period, the roadway surface displacement presents the characteristic of "convergence on two sides > floor heave > vault settlement". Up to the 120th day of monitoring, the deformation still continues to increase with the total convergence

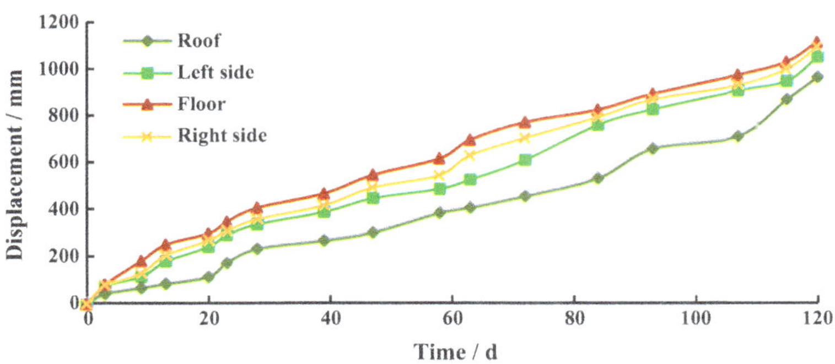

Fig. 6.4 Displacement curves of roadway surface

of 2155 mm on the two sides; and the floor heave has reached to 1120 mm and the vault settlement has reached to 967 mm.

2. Monitoring on the Destruction Range of Surrounding Rock

A borehole television is used to monitor the destruction range. In order to analyze the internal destruction of the roadway surrounding rock, monitoring sections are arranged between each section on the typical roadway area. 5 detecting boreholes of Φ42 mm are set on each section on the vault, the left and right spandrel, and the left and right sides of the roadway, respectively. To facilitate the result analysis on each detecting borehole, the interpretation standard of borehole detection is established for Liangjia Coal Mine; and it is also for classifying the integrity and the destruction range of roadway surrounding rocks. As shown in Fig. 6.5, the black filings of each borehole represent the observed destruction positions and ranges on the borehole inner wall. According to the diagram of the detection results, the destruction range of surrounding rock is distinguished as severe, medium and slight damage zones from inside to outside.

The analysis shows:

(1) The surrounding rock destruction is large in range and deep inside. The slight damage zone reaches to 6000 mm of the detecting borehole; the outside boundary of the medium damage zone is 3900 mm averagely away from the chamber wall (about 1.4 times of the chamber radius); and the severe damage zone is 2800 mm averagely away from the chamber wall (about 1.1 times of the chamber radius).

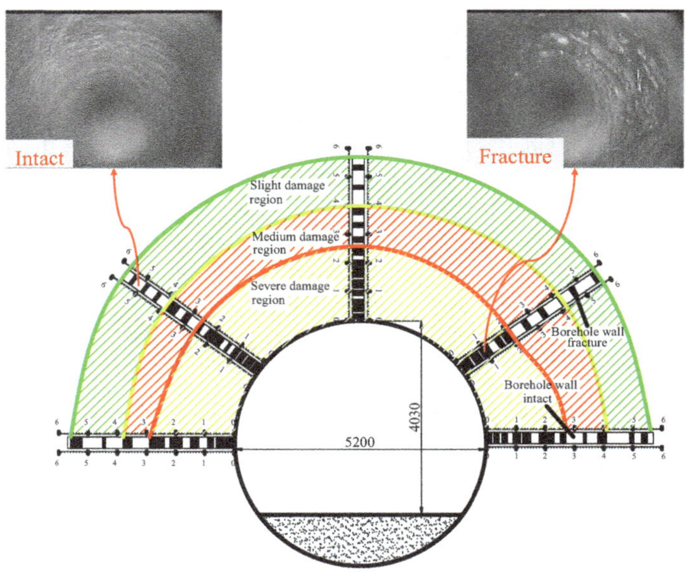

Fig. 6.5 Damage range of surrounding rock detection result (mm)

(2) The destruction range around the roadway is in an approximately identical ring shape as the roadway shape. In the severe damage zone, surrounding rocks are crushed and cracks are extremely developed. In the borehole, the proportion of the total number of cracks can be clearly distinguished as "on the vault > at the right side > at the left side"; and the proportion is larger in the shallow surrounding rock than in the deep surrounding rock.

6.2.2 Failure Mechanism of Supporting Components

1. Monitoring on Bolt Stress

Bolt dynamometers are arranged on the selected typical monitoring sections and located on the left side, the vault and the right side, respectively. Figure 6.6 shows the monitoring results.

The analysis shows the severely crushed surrounding rock has the destruction range far beyond the bolt length; and therefore, it is unable to provide the stable anchor foundation for bolts. As a result, the bolt has the low initial pre-tightening force and the small post force with the maximum value of only 33 kN. Moreover, the bolt is anchored with a resin of 600 mm in length, and its insufficient anchorage length results in its insufficient anchorage force in the original supporting scheme. In addition, the surrounding rock in the borehole is easy to collapse; and the rock is easy to fall off from the borehole on the field construction of the bolt support. As a result, the anchor agent fails to make dense filling between bolts and surrounding rock; so the bolt cannot effectively play its support potential with its anchorage force further reduced.

2. Monitoring on Arch Stress

Pressure gauges are used to monitor radial force on the U36 arch. The monitoring section is arranged with 5 monitoring points which are located on the left and right

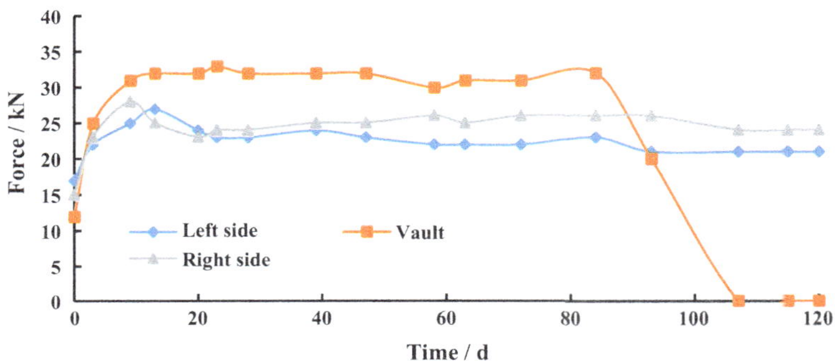

Fig. 6.6 Bolt force curves

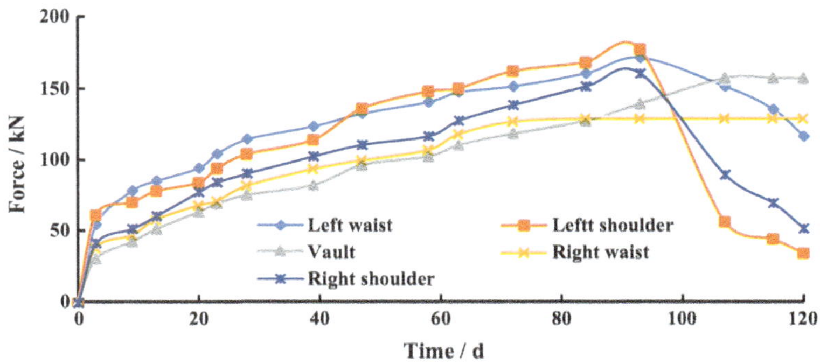

Fig. 6.7 Arch force curves

hances, the left and right spandrels, and the vault of U36 arch, respectively. Figure 6.7 shows the monitoring results.

The analysis shows:

(1) Within 1–3 days after the pressure gauges being installed, the force has reached 60 kN at some points. Until the 63th day of monitoring, the force has been rising on each point over time, especially the radial force on the left spandrel. The radial force reaches 151 kN on the 63th day and its maximum of 179 kN on the 93th day. It starts to decline on some points after the 93th day and declines to 35 kN at the point on the left spandrel on the 120th day. The arches are clearly buckling, some are failure, and basically lose their bearing capacity.
(2) Wood boards and other special means are used during construction to achieve the uniform contact among the pressure gauge, the arch and the surrounding rock, and to better play the bearing capacity of the arch. The monitoring data shows that the total radial force of the arch is 780 kN at most; and other arches in the normal construction are broken, as shown in Fig. 6.3. All of that indicate the U36 arch has insufficient bearing capacity, low strength, and uneven contact with surrounding rock; and it cannot meet the requirement of stable control on surrounding rock under this geological condition.

The following conclusions are made from the field monitoring of deep soft rock roadways:

(1) Under the traditional support, the roadway surface displacement presents the characteristic of "convergence on two sides > floor heave > vault settlement". The destruction range of surrounding rock is large and the severe destruction zone is 2800 mm averagely away from the chamber wall.
(2) The bearing capacity of the U36 steel arch is insufficient, which cannot meet the requirements of stability control of surrounding rock. Anchor bolts are generally in the severe destruction zone; their support potential fail to be played.

6.3 Control Mechanism of Soft Rock Roadways

6.3.1 Design of Numerical Tests

The main factors that influence the stability of surrounding rock in roadways are supporting strength, geo-stress, mechanical parameters of surrounding rock and others. Considering the production requirement on-site and the limitation of anchor bolts, in this section, numerical models are established to study the mechanical properties of the support components and the surrounding rock control mechanism under the influence of those factors. The size of the roadway cross-section on site and the supporting parameters of the bolt-shotcrete are used as invariants; and arch strength, geo-stress and mechanical parameters of surrounding rock are used as variables. The numerical comparative test schemes are designed in 3 categories.

6.3.1.1 Comparative Schemes of Arch Strength

The geo-stress and the mechanical parameters of surrounding rock measured from downhill track at No. 6 Mining Areas of Liangjia Coal Mine are used as invariants; and the arch strength is used as variables. The comparative analysis is made on the influence law of the arch strength on the bearing mechanism of roadway support. The support is in the form of bolt-shotcrete and ($i/4$) U36 arch. 9 schemes are designed with the corresponding serial number of A_i. The specific schemes are shown in Table 6.1, where, $i/4$ represents arch strength grade, and $i = 0 - 8$ corresponds to 0–2 times of the strength of the U36 arch, respectively.

6.3.1.2 Comparative Schemes of Geo-Stress

The support is in the form of bolt-shotcrete + U36 arch with the measured mechanical parameters of surrounding rock as invariants and geo-stress as variables. 7 schemes are designed with corresponding serial number of B_j to analyze the influence law of geo-stress on the bearing mechanism of roadway support. The specific schemes are shown in Table 6.2, where, $j = 1 - 7$ corresponds to 0.25–1.75 times of the measured geo-stress, respectively. B4 represents measured geo-stress.

Table 6.1 Comparison scheme of arch strength

Scheme no.	Arch strength grade	Supporting scheme	Invariant	Remark
A_i	$i/4$	Bolt-shotcrete + ($i/4$) U36 arch	Measured geo-stress Measured mechanical parameters of surrounding rock	$i = 0 \sim 8$

Table 6.2 Comparison schemes of geo-stress

Scheme no.	Geo-stress grade	Geo-stress/MPa	Invariant
B1	0.25	5.5	Bolt-shotcrete + U36 arch with the measured mechanical parameters of surrounding rock
B2	0.5	11	
B3	0.75	16.5	
B4	1.0	22	
B5	1.25	27.5	
B6	1.5	33	
B7	1.75	38.5	

Table 6.3 Comparison schemes of mechanical parameters of surrounding rock

Scheme no.	Mechanical parameters grade	Elastic modulus E/GPa	Cohesion c/MPa	Angle of internal friction $\varphi/°$	Invariant
C1	0.4	0.52	0.162	15.2	Bolt-shotcrete + U36 arch measured geo-stress
C2	0.6	0.78	0.189	17.6	
C3	0.8	1.04	0.216	20.16	
C4	1.0	1.3	0.27	25.2	
C5	1.2	1.56	0.34	32.5	
C6	1.4	1.82	0.41	37.8	
C7	1.6	2.08	0.47	44	

6.3.1.3 Comparative Schemes of Mechanical Parameters of Surrounding Rock

The support is in the form of bolt-shotcrete + U36 arch with the measured geo-stress as invariant and mechanical parameters of surrounding rock as variables. Elastic modulus E, cohesion c and the angle of internal friction φ are mainly taken as the reference standard of mechanical parameters of surrounding rock. 7 schemes are designed with C_m as corresponding serial numbers to analyze the influence law of the mechanical parameters of surrounding rock on the bearing mechanism of the roadway support. The specific schemes are shown in Table 6.3, where, $m = 1 - 7$, corresponds to 0.4–1.6 times of the measured mechanical parameters of surrounding rock, respectively. C4 represents measured mechanical parameters of surrounding rock.

6.3.1.4 Modeling Establishment and Structural Unit Settings

1. Modeling Establishment

The models are simplified to obtain the law with more universal significance and clarify the supporting bearing mechanism. The strata are simplified as level and the surrounding rock is not distinguished by rock lithology. The lateral pressure coefficient is set to 1, and a 1/4 circular chamber is taken for modeling.

The model dimension is 20000 mm × 20000 mm × 16000 mm (width × height × thickness). The model uses hexahedron elements with a total of 8528 units and 11158 nodes. The model is shown in Fig. 6.8.

2. Structural Unit Settings and Parameters

The cross-section of the roadway is one-quarter circular with the radius of 2700 mm and the roadway excavation is completed by one time. After the completion of the excavation, the shotcrete are sprayed, the anchor bolts and the arches are installed. The concrete spray layer is cling to the excavation surface of the roadway with the thickness of 100 mm.

The model is arranged with 2 rows of bolts with cable unit; the inter-row spacing is 650 × 800 mm; and the mechanical parameters are shown in Table 6.4.

The model is arranged with 2 rows of arches symmetrically with beam unit; the row spacing of arches is 650 × 800 mm. The arches and bolts are arranged in the same section. The parameters of the arch and the backplane are shown in Table 6.5.

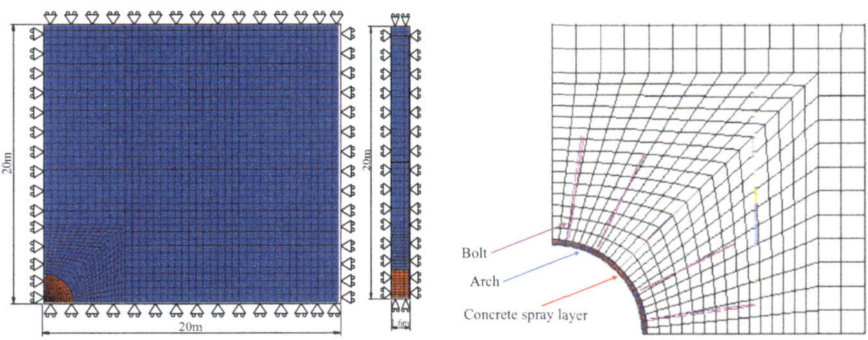

(a) The overall view of model (b) The partial view of model

Fig. 6.8 Numerical model

Table 6.4 Bolt parameters

Support components	Components size/mm	Poisson ratio μ	Yield strength σ_s/MPa	Ultimate strength σ_b/MPa	Elastic modulus E/GPa
Bolt	Φ25 × 2250	0.3	500	700	200

Table 6.5 Arch parameters

Components formation	Elastic modulus E/GPa	Sectional area S/cm^2	Moment of inertia I_x/m^4	Moment of inertia I_y/m^4
U36	206	45.69	9.29×10^{-6}	1.25×10^{-5}

6.3.2 *Result Analysis*

The comparative analysis is made on the deformation of the surrounding rock, the range of the plastic zone and the stress of anchor bolts. The supporting bearing mechanism of deep roadways with soft rock is clarified through the analysis. In the analysis, the deformation of the surrounding rock refers to the largest deformation D_{m}. The plastic range refers to the plastic damage area of the surrounding rock R_{p}, which represented by the equivalent radius and obtained through volume conversion of the statistical plastic area. The stress of anchor bolts refers to the average value of the maximum axial force F_{n} of all bolts in a numerical model.

6.3.2.1 Result Analysis of Comparative Schemes of Arch Strength

The calculation results of comparative schemes of arch strength are plotted in Figs. 6.9 and 6.10. The analysis shows:

(1) In scheme A0, with support of bolt-shotcrete, the displacement of surrounding rock reaches to 1524 mm; the range of the plastic zone reaches to 10300 mm. The surrounding rock is seriously deformed and broken, and therefore, it is unable to form the stable and reliable bearing structure of the surrounding rock.
(2) With the combined support of bolt-shotcrete + U36 arch, the displacement of surrounding rock decreased to 725 mm; and the displacement is 52.4% smaller than that with the bolt-shotcrete support. The range of the plastic zone decreased to 5520 mm; and it is 46.4% smaller than that with the bolt-shotcrete support. The

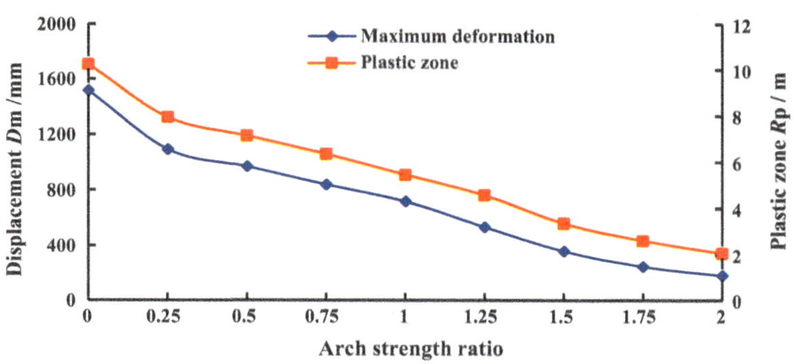

Fig. 6.9 Displacement and plastic zone curve of surrounding rock with different arch strength

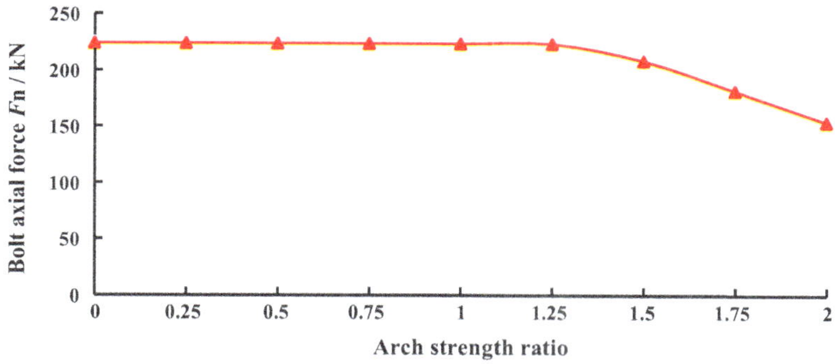

Fig. 6.10 Bolt axial force curve of different arch strength

data indicates that the U36 arch has remarkable control effect on the surrounding rock. However, the surrounding rock is still in the state of serious deformation and destruction. In schemes A2–A8, both displacement of the surrounding rock and range of the plastic zone are in the trend of approximate linear decrease along with the increase of arch strength.

(3) In this example, the combined support of bolt-shotcrete + U36 arch is unable to provide the effective control on the stability of deep roadways with soft rock. The bolts are exposed to the plastic zone of the surrounding rock; the axial force reaches the breaking load; and the bearing structure of the surrounding rock is unstable. This situation is basically consistent with the field test conclusion.

(4) When the arch strength grade is less than 1.25, the axial force of all bolts reaches the breaking load; and when the arch strength grade is greater than 1.25, the axial force of bolts gradually decreases along with the increase of arch strength. That indicates that no matter with or without the support of low strength arch, the stress of the surrounding rock on bolts is too large and will lead to break of the bolt easily; in the other hand, with the support of high strength arch, the stress of the surrounding rock on bolts is small and the bolts has not reached the ultimate load, and the bolt has the good supporting strength reserve.

In summary, the support strength of bolt-shotcrete is very limited for deep roadways with soft rock. Its insufficient strength results in the large deformation and serious plastic damage of the surrounding rock, and its support components exposed to their breaking load. It is hard for the surrounding rock to form the effective and stable bearing structure. However, those roadway support problems could be solved effectively by increasing the arch strength. Both displacement of the surrounding rock and the range of the plastic zone are in the significantly decreasing trend along with the increase of the arch strength.

6.3.2.2 Result Analysis on Comparative Schemes of Geo-Stress

The calculated results of comparative schemes of geo-stress are plotted in Figs. 6.11 and 6.12.

The analysis shows:

(1) Along with the increase of geo-stress, the displacement of the surrounding rock is in the substantially exponential growth trend; thus, the geo-stress has the significant influence on the surrounding rock deformation.
(2) Along with the increase of geo-stress, the range of plastic zone is approximately in the logarithmic growth trend. When the geo-stress is greater than 22 MPa, the growth rate of the range of plastic zone is reduced and the range of plastic zone gradually approaches to 7200 mm.
(3) The axial force of bolts is approximately in the logarithmic growth trend along with increase of geo-stress. When the geo-stress is greater than 16.5 MPa, the

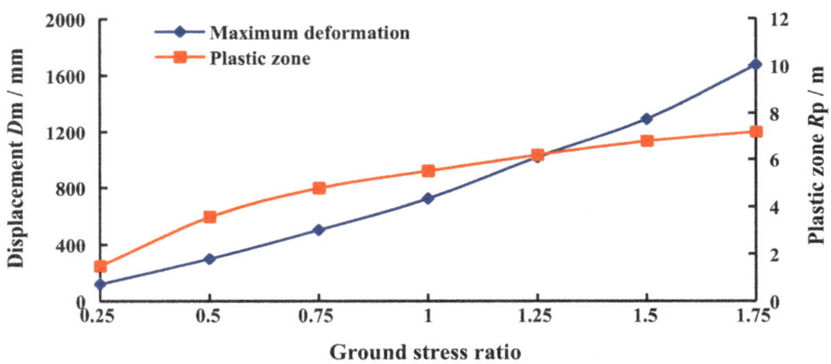

Fig. 6.11 Displacement and plastic zone curve of surrounding rock with different geo-stress

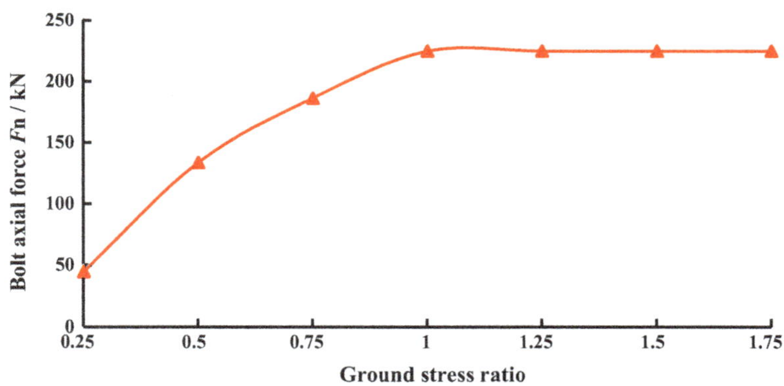

Fig. 6.12 Bolt axial force curve of different geo-stress

growth rate of the axial force of bolts slows down and the force reaches to the breaking load.

In summary, the increase of geo-stress results in the very large deformation of the surrounding rock and the serious plastic damages, and the bolt reaches or closes to its breaking load. In this example, the combined support of bolt-shotcrete + U36 arch is hard to meet the requirement of stable control on surrounding rock when the geo-stress is greater than 22 MPa, the bearing structure is unable to bear the pressure of surrounding rock and will end up in failure.

6.3.2.3 Result Analysis on the Comparative Schemes of Mechanical Parameters of Surrounding Rock

The calculated results of the comparative schemes of mechanical parameters of surrounding rock are plotted in Figs. 6.13 and 6.14.

The analysis shows:

(1) Along with the increase of mechanical parameters of surrounding rock, the displacement of the surrounding rock is in the exponential decrease trend. When the mechanical parameter grade is less than 1.2, the displacement of the surrounding rock is severely influenced by the mechanical parameters of surrounding rock; and the deformation in Scheme C1 is 1622 mm, which is 44.7 and 26.6% of that in Schemes C4 and C5. When the mechanical parameters grade is greater than 1.2, the influence weakened on the surrounding rock deformation.

(2) Along with the increase of mechanical parameters of surrounding rock, the range of plastic zone is in the exponential decrease trend. The range of plastic zone is 8870 mm in Scheme C1 and 2800 mm in Scheme C6, and the latter

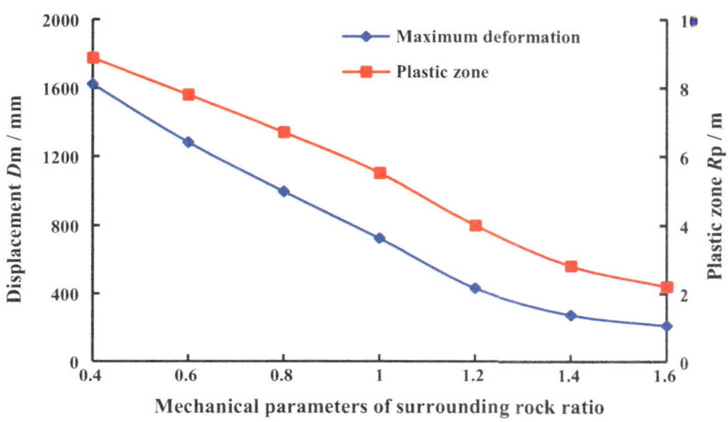

Fig. 6.13 Displacement and plastic zone curve of surrounding rock with different mechanical parameters

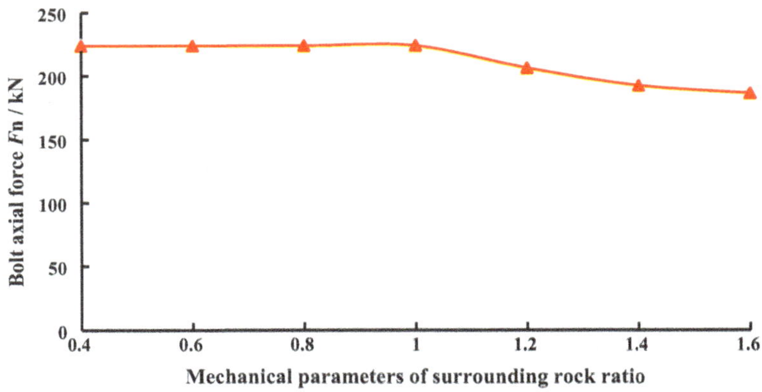

Fig. 6.14 Bolt axial force curve of surrounding rock with different mechanical parameters

reduced by 68.4%. The influence of mechanical parameters of surrounding rock is significant in the plastic zone.

(3) When the mechanical parameter grade is less than 1, the stress on all bolts reaches their breaking load; and as the mechanical parameter grade is greater than 1, the bolt axial force is in the linearly decreasing trend. The bolt axial force is reduced by 14.2 and 16.9% of the breaking load in Schemes C6 and C7.

In summary, with the higher grade of mechanical parameters of surrounding rock, the traditional support can meet the requirements of stable control on surrounding rock. In the other hand, with the lower grade of mechanical parameters of surrounding rock, both surrounding rock deformation and the range of plastic zone increased; the support components are close to failure; and the surrounding rock control becomes more difficult. In this example, when the mechanical parameters grade of surrounding rock is less than 1, the traditional combined support of bolt-shotcrete + U36 arch is hard to meet the requirement of stable control on surrounding rock; and the bearing structure of the surrounding rock is unable to bear the pressure of surrounding rock effectively, resulting in the failure of the bearing structure.

The following conclusions are made from the numerical test:

(1) The bolt-shotcrete support is very limited in strength on the deep roadway with soft rock. The insufficient strength results in the large deformation of surrounding rock and the supporting components reaching their breaking load. The surrounding rock is difficult to form the stable and effective self-bearing structure.

(2) With the increase of geo-stress, the strength of surrounding rock decreases, and the deformation of surrounding rock, the range of plastic zone and the stress of supporting components all show the increasing trend. The traditional high strength support system (bolt-shotcrete + U36 arch) cannot meet the stable control requirements of surrounding rock.

(3) With the increase of the strength of the arch, the deformation of the surrounding rock and the range of plastic zone show the significant decrease trend, and the surrounding rock control effect is significantly improved. Therefore, the support with high strength arches is effective to solve the surrounding rock control problems of this type of roadways. At the same time, a certain amount of pressure yielding should be ensured on the basis of the high strength support, and the self-bearing capacity of surrounding rock should be fully exerted.

6.4 Engineering Application of New High Strength Support

6.4.1 Field Application of CCC Support

6.4.1.1 Application Location

The #6 Mining Areas is located in the northeast of the Liangjia Coal Mine Field with its mining depth of −800 m. In this chapter, the downhill track of the #6 Mining Areas is selected to conduct the research, and the location of the mining area is shown in Fig. 6.15. The stratum structure of the roadway is extremely complicated and unstable. Composed of carbonaceous mudstone, sandy mudstone and mudstone-clayed rock, the direct roof is easy to be weathered and fall off; and it absorbs water and expands, and belongs to the easy-falling roof. The direct floor is basically composed of mudstone, and the part of it is carbonaceous mudstone, oil-bearing mudstone and oil shale; and it is easy to absorb water and expand. Besides, the floor heave of the roadway is severe. As an unstable roadway with extremely soft rock, it is very difficult to control.

The roadway cross-section is round with the original support of bolt-shotcrete + U36 steel arches, as shown in Fig. 6.16. The bolt type is Φ25 × 2250 mm and the row spacing is 650 × 800 mm; the surface of the roadway is sprayed with C20 concrete with the thickness of 100 mm; and the arch is made of heavy-duty U36 steel with the row spacing of 800 mm.

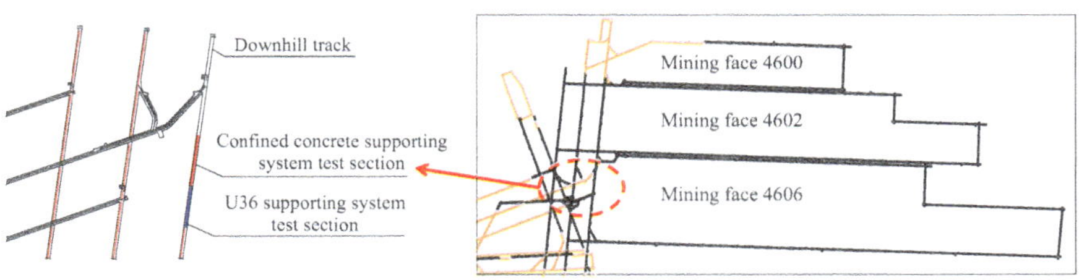

Fig. 6.15 Location of tested roadway

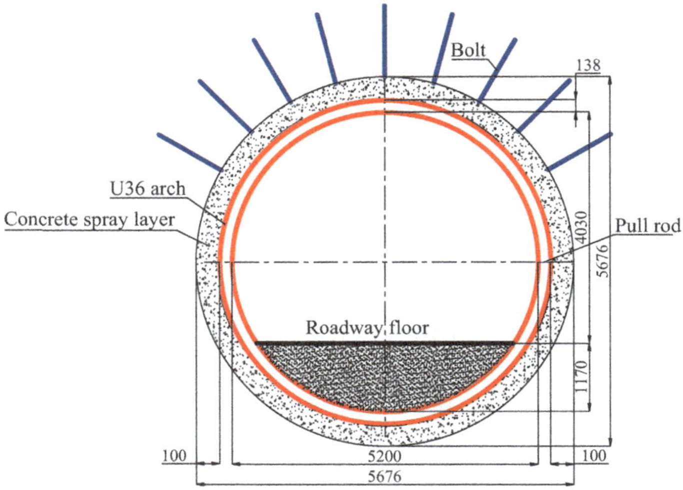

Fig. 6.16 Original supporting program (mm)

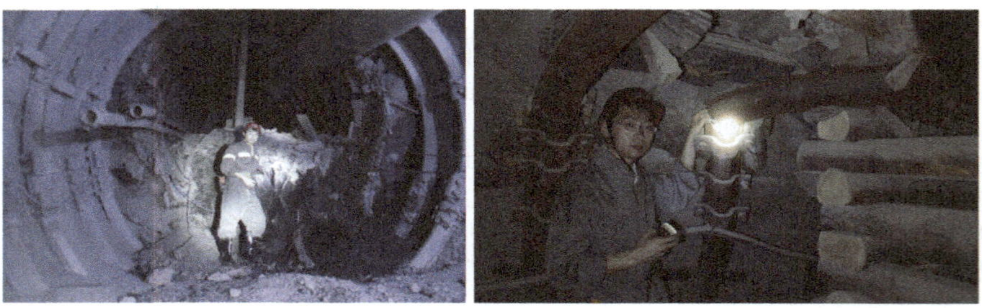

Fig. 6.17 Roadway destruction

The large deformation of surrounding rock and broken arches still occur even under such the strong combined support, as shown in Fig. 6.17.

6.4.1.2 Scheme Design

The tested cross-section size of the CCC support system is consistent with the original design of the roadway. The supporting cross-section is shown in Fig. 6.18. The bolt arrangement and the spray layer are consistent with the original support scheme.

(1) The CCC arch is divided into five sections, which are connected by casings. The casing size is 194 mm × 14 mm (diameter × wall thickness). Quantitative pressure yielding devices are installed in the four casings beside the casing on the bottom.

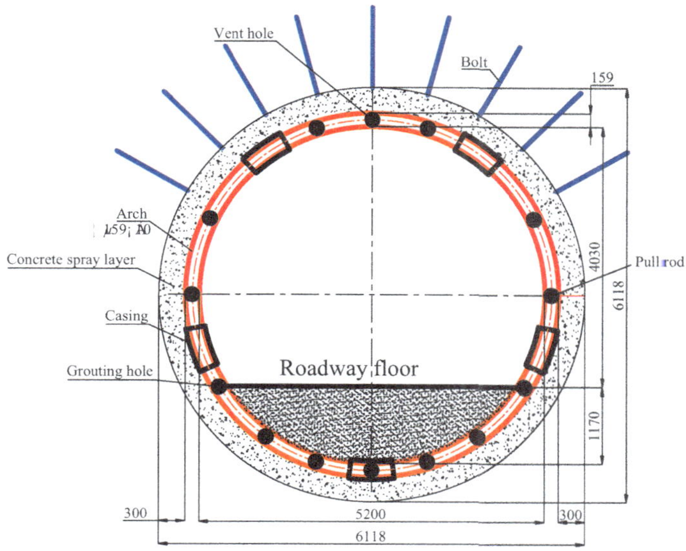

Fig. 6.18 Size of arch in the tested roadway (mm)

(2) The cross-section of the arch is round, the size of the steel tube is 159 × 10 mm (diameter × wall thickness); and the inner diameter (radius) is 2600 mm, and the diameter is 5200 mm. The arch spacing is designed to be 800 mm. The arches are connected with ejector pins; and the distance between the ejector pins is 1500 to 1800 mm, which can effectively prevent the arch from losing their stability.

(3) After the roadway is enlarged, the initial injection is made with anchor bolts (cables) and arches installed immediately. After the arch is assembled, the concrete core is poured. The steel tube is filled with C40 concrete. Each section of the arch is arranged with a grouting hole and a venting hole. The grouting hole diameter is designed as 80 mm, and side bending steel plates are welded on both sides of the grouting hole for reinforcement.

6.4.1.3 Field Implementation

(1) After the blasting excavation, the bolt-shotcrete is first carried out.

(2) The arch is assembled from bottom to top. Firstly, the arch bottom is set up; then the arch sections are connected with the casing joints at the two sides and laterally fixed with bolts; and finally, the vault is set up. A whole arch is composed of five arch sections and fixed on the roof with bolts. The assembly of arch frame is completed.

(a) Installation of two sides of the arch

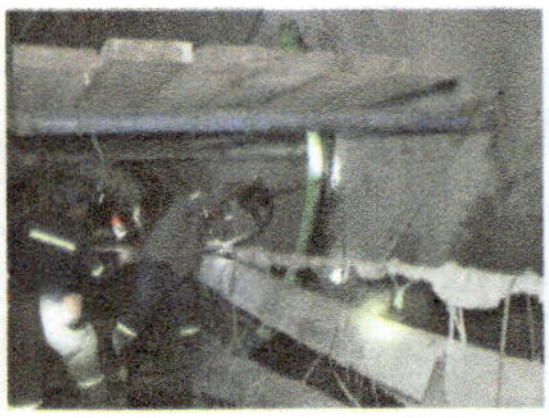

(b) Installation of the tilting preventing device

(c) Installation of the casing joint

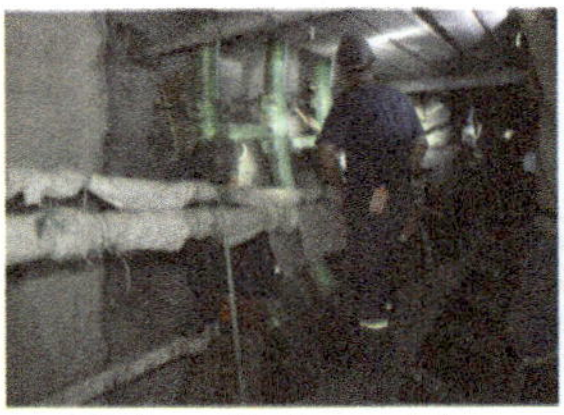

(d) Installation of the arch sides

(e) Installation of the vault

(f) Concrete core pouring

Fig. 6.19 Implementation of the CCC arch support scheme in field

(3) After the arch support is completed, the metal mesh back plate is laid, the back wall is filled and finally the concrete core is poured.
(4) After the completion of the support, follow-up monitoring will be carried out. The field implementation is shown in Fig. 6.19.

6.4.1.4 Monitoring Comparison

Long-term follow-up monitoring is carried out on the surrounding rock convergence of the roadway in the CCC arch application area after the arch erection. The deformation monitoring results in the tested section with the confined concrete and the U36 arch support are shown in Fig. 6.20.

The deformation of the surrounding rock in both of two test sections show the phenomenon of "floor heave > deformation at the sides > roof settlement". At the 120th day of monitoring, the average deformation of each measuring point is 424 mm; and the average deformation is 95 mm at each measuring point in the tested section with confined concrete arch support, which is 22.4% of that with the U36 arch support. The monitoring results in Fig. 6.20 show the surrounding rock of the tested section with U36 arch support deforms continuously and cannot reach the stable state. However, the surrounding rock deformation is basically in the stable state in the tested section with confined concrete support after the 84th day of monitoring, and the surrounding rock is effectively controlled.

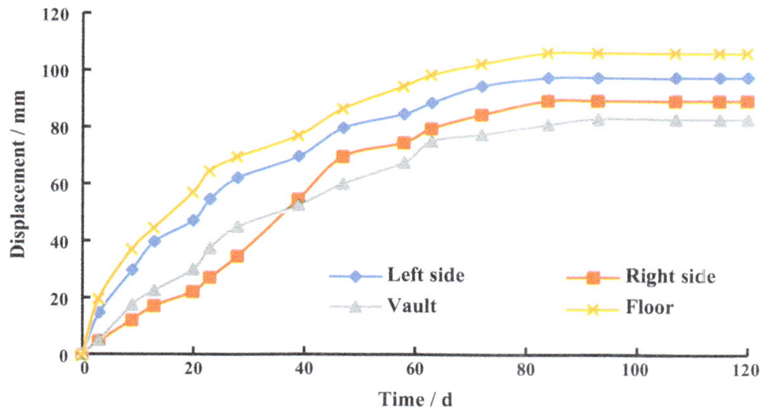

(a) Deformation in the tested section with the confined concrete arch support

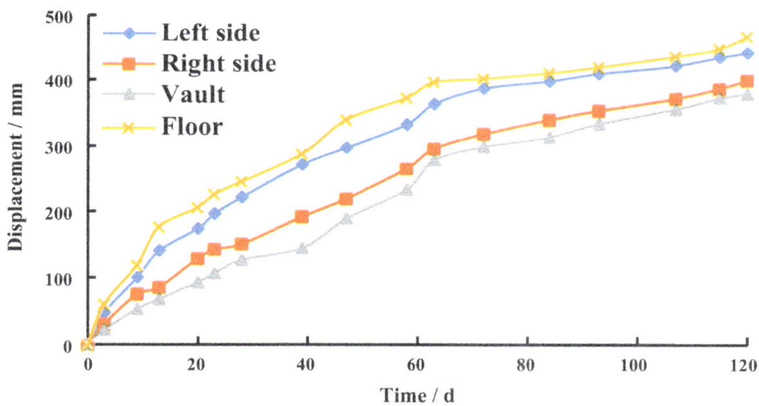

(b) Deformation in the tested section with the U36 arch support

Fig. 6.20 Displacement curves of roadway surface

The surrounding rock control effect in the tested section with confined concrete support is shown in Fig. 6.21.

6.4.2 Field Application of SQCC Support

6.4.2.1 Application Location

In this section, the track hidden inclined shaft of Coal 4 in #6 Mining Areas is selected for the research. The structure of the stratum is extremely complicated with the poor stability; and the dirt band is mainly composed of mudstone, then carbonaceous

Fig. 6.21 Surrounding rock control effect of CCC arches

mudstone with uncertain thickness, as shown in Fig. 6.22, The direct roof is mainly composed of carbonaceous mudstone, sandy mudstone and mudstone-clay rock, which is easy to be weathered and fall off, absorbs water and expands. The bottom is 0.64 m mudstone and partially carbonaceous mudstone, oil-bearing mudstone and oil shale. It is easy to absorb water and expand with the severe floor heave at the working face. The roadway has the characteristic of typical three-soft with poor stability.

The cross-section of the roadway is round with the original support of bolt-shotcrete + U36 steel arch, as shown in Fig. 6.23. The bolt size is Φ25 × 2250 mm,

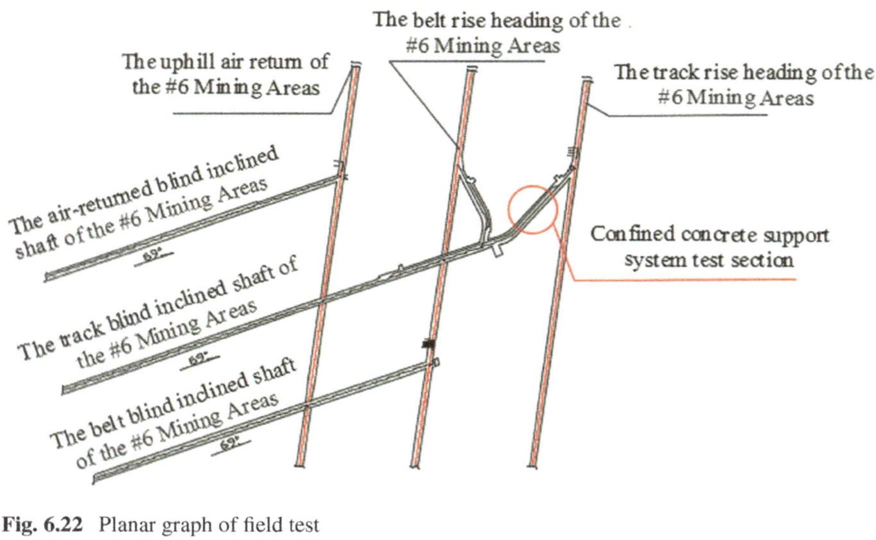

Fig. 6.22 Planar graph of field test

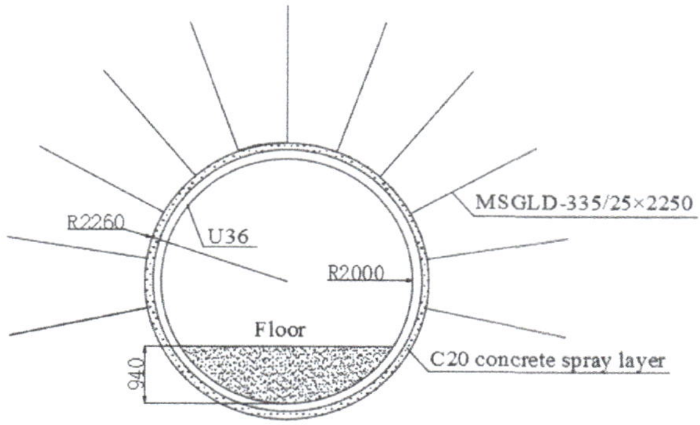

Fig. 6.23 Section design drawing of original supporting scheme (mm)

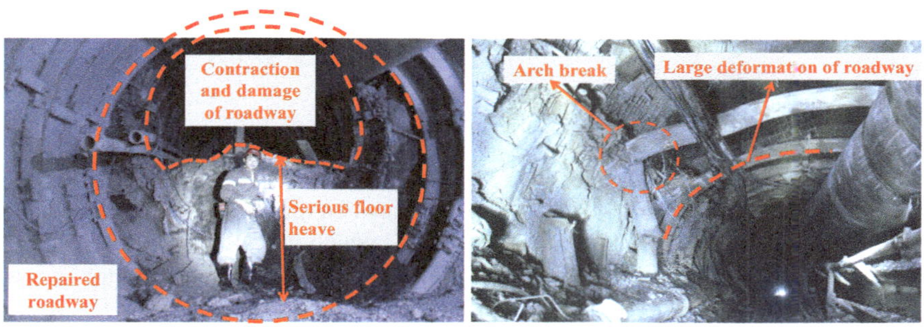

Fig. 6.24 Field failure situation of original supporting scheme

the row spacing is 800 × 800 mm; the surface of the roadway is sprayed with C20 concrete with the thickness of 100–120 mm; and the arch is made of the heavy-duty U36 steel with the row spacing of 800 mm.

In the original support scheme, many U36 arches are buckled and broken. The large deformation and severe floor heave occur on the roadway, as shown in Fig. 6.24. The support strength needs to be greatly improved on the roadway.

6.4.2.2 Scheme Design

The cross-section size of the tested section of the SQCC arch is consistent with the original design of the roadway. The supporting cross-section is shown in Fig. 6.25. The bolt arrangement and the spray layer are consistent with the original support scheme.

(1) The SQCC arch is divided into five sections with flanged connection.
(2) The cross-section form of the arch is square; the size of the steel tube is 150 × 8 mm; the inner diameter of the arch is 4000 mm; and the spacing of the arch

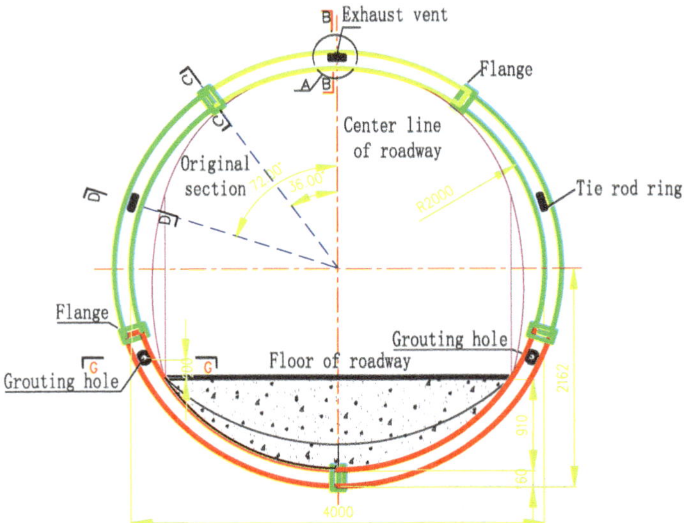

Fig. 6.25 Sectional drawing of SQCC arch support (mm)

is 800 mm. The ejector pins are installed between the arches; and the distance between the ejector pins is 1500–1800 mm, which can effectively prevent the arch from losing their stability. According to the theoretical calculation, a row of bolts is installed horizontally at each side for reinforcement.

(3) After the roadway is enlarged, the initial injection is made with the bolts (cables) and arches installed. After the arches are assembled, the concrete core is poured. The concrete is C40. Each section of the arch has a grouting hole and a venting hole. The diameter of the grouting hole is designed to be 80 mm, and the side bend plate is welded on both sides of the grouting hole for reinforcement.

6.4.2.3 Field Implementation

The construction process of the SQCC support is:

(1) After the blasting excavation, the bolt-shotcrete is first carried out.
(2) The arch is assembled from bottom to top. Firstly, the arch bottom is set up; then the arch sections are connected with the flanged connection at the two sides and laterally fixed with bolts; and finally, the vault is set up. A whole arch is composed of five arch sections and fixed on the roof with bolts. The assembly of arch is completed.
(3) The metal mesh back plate is laid, the back wall is filled and finally the concrete core is poured.
(4) After the completion of the support, follow-up monitoring will be carried out. The field implementation is shown in Fig. 6.26.

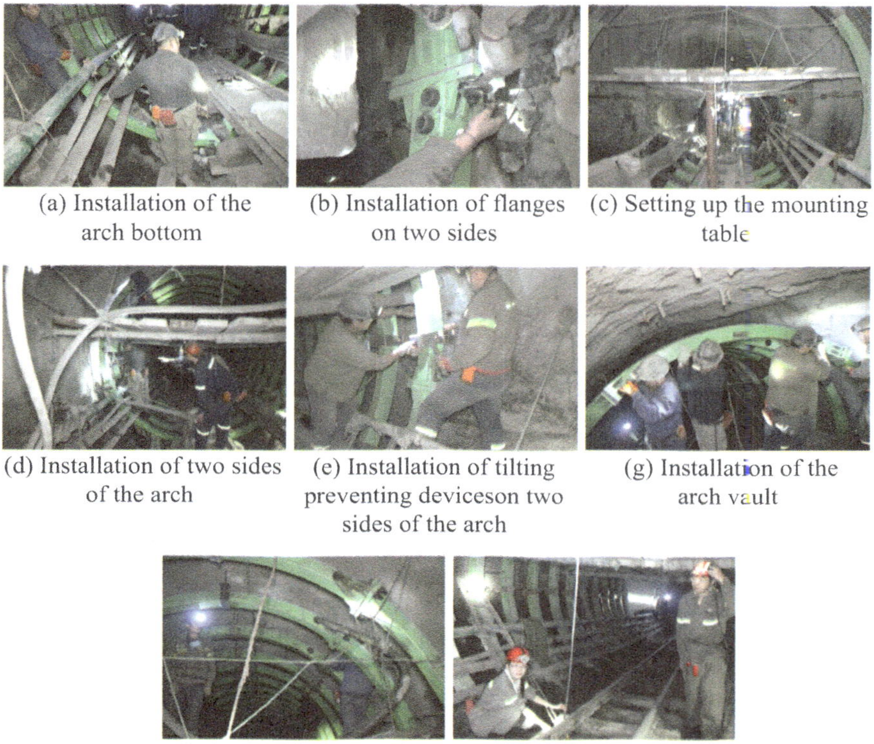

(a) Installation of the arch bottom

(b) Installation of flanges on two sides

(c) Setting up the mounting table

(d) Installation of two sides of the arch

(e) Installation of tilting preventing deviceson two sides of the arch

(g) Installation of the arch vault

(h) Concrete core pouring

(i) Construction quality monitoring and inspection

Fig. 6.26 Implementation of the SQCC arch support scheme in field

6.4.2.4 Monitoring Comparison

Long-term follow-up monitoring is carried out on the surrounding rock convergence of the roadway in the SQCC arch application area after the arch erection. The monitoring results are shown in Fig. 6.27.

According to on-site monitoring, the results show the roadway convergence deformation at right side > at left side > on right spandrel > on vault > on left spandrel; the deformation is faster in the first 15 days and slows down about 25 days; and after 75 days, the roadway is basically stable and has almost no further deformation; the average deformation of the measuring points is 26 mm; and the SQCC arch support has a better control effect on surrounding rock.

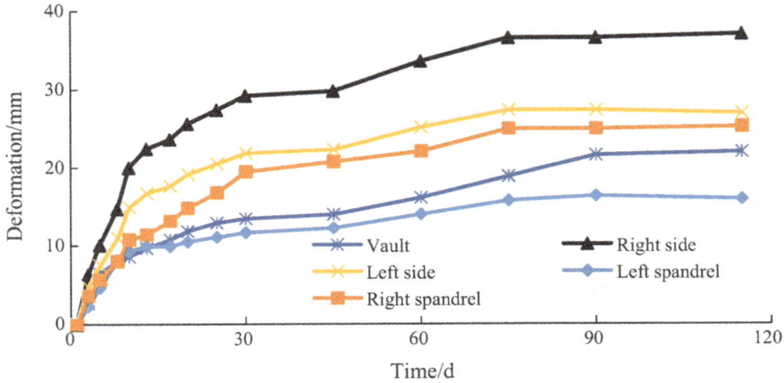

Fig. 6.27 Displacement curves of roadway surface

Fig. 6.28 Surrounding rock control effect of SQCC arches support

The control effect of surrounding rock in the tested section with confined concrete arches is shown in Fig. 6.28.

6.5 Chapter Summary

In this chapter, the Liangjia Coal Mine, the extremely soft rock roadway in the sea area, is used as the engineering background. Field monitoring is carried out on the convergence deformation and the damage range of the surrounding rock, and the stress on the supporting components. With the monitoring, the failure of the bearing structure of the typical deep soft rock roadway and the surrounding rock control mechanism are clarified. Through numerical analysis on surrounding rock deformation, plastic zone range and bolt stress, the supporting mechanism of deep soft rock roadways are defined. The field test shows that the surrounding rock is effectively controlled in the tested roadway sections with CCC and SQCC arches, verifying the confined concrete support has a good control effect on surrounding rock.

Chapter 7
Engineering Practice of New High Strength Support System in Deep Roadways with High Stress

In this chapter, Zhaolou Coal Mine, a typical kilometer deep mine is taken as the engineering background. A full scale laboratory comparison experiment is made on the SQCC arch and the traditional U29 arch to establish the quantitative evaluation index. The comparative analysis is made under different factors on the surrounding rock control mechanism and the mechanical properties of the supporting components as well. Combining with the comparison test on-site, our study verifies that the SQCC support is very effective on the surrounding rock control. This research could provide references for solving the support problem of roadways with high stress.

7.1 Project Overview

As a large coal field, Juye Coal field is located in the Southwest of Shandong Province of Eastern China. With the burial depth of 800–1300 m, its main coal seam has the average thickness of 8 m of No. 3 coal. The total proven geological reserves are about 5.57 billion tons. Zhaolou Coal Mine is located in the middle of Juye coal field, as shown in Fig. 7.1. Its designed production capacity is 3.0 Mt/a.

In this paper, the second set of track downhill is selected as the research object, as shown in Fig. 7.2. Its roadway is located in an area of tectonic development and runs through the DF7 fault with the break distance reaching more than 10 m. Fractures are developed in the fault crushing zone. The surrounding rock of the roadway is mainly mudstone with some coal seam exposed. With the very poor quality, the rock is hard to be controlled.

The section of the roadway is in the shape of straight wall and semicircular with the net width of 5000 mm and the net height of 4300 mm; and its support form is a combination of bolt-shotcrete + U29 arches, as shown in Fig. 7.3. The bolt is the model Φ22 × 2400 mm with row and line spacing of 800 × 800 mm; the cable is the model Φ22 × 6200 mm with row and line spacing of 2000 × 1600 mm; C25

Q. Wang et al., *High Strength Support for Soft Surrounding Rock in Deep Underground Engineering*, https://doi.org/10.1007/978-981-15-3844-5_7

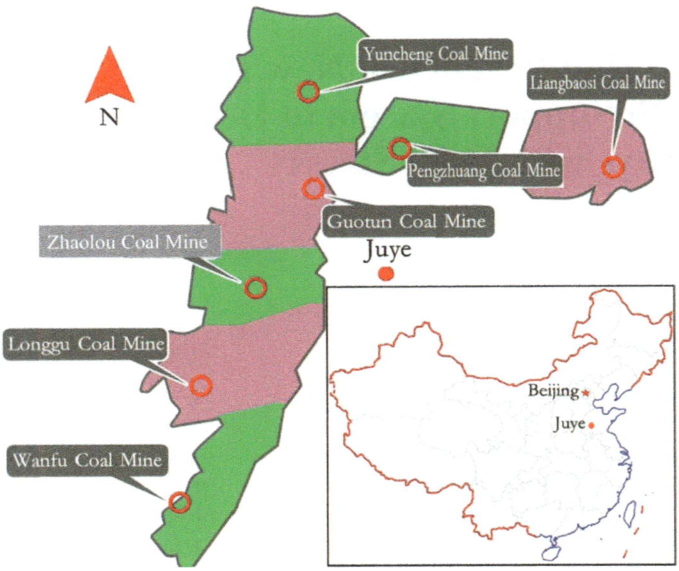

Fig. 7.1 Location of Juye coal mine

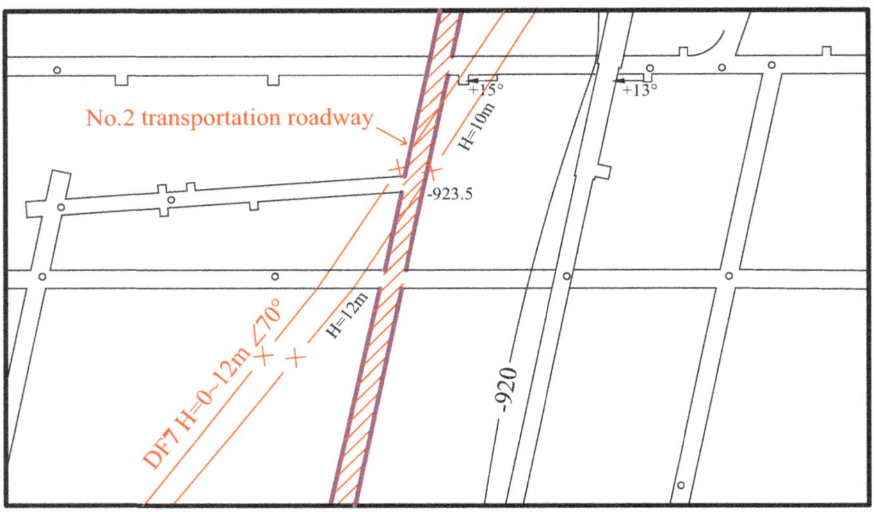

Fig. 7.2 Location of research roadway

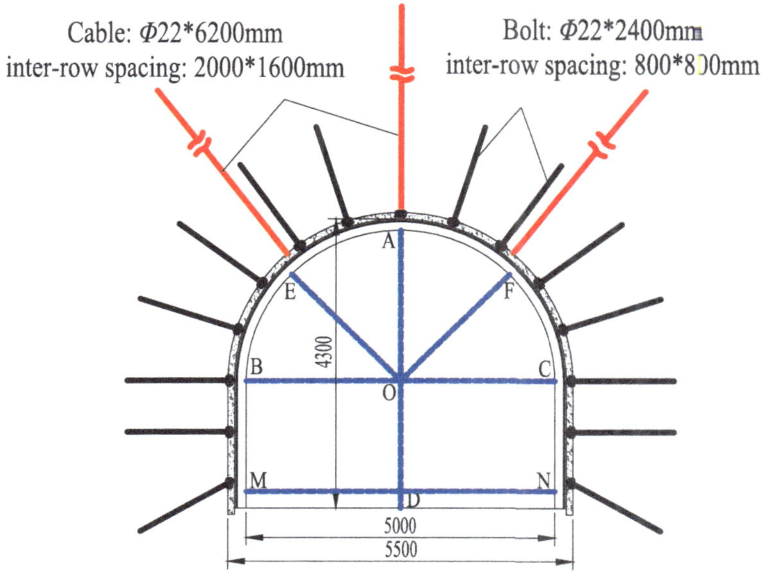

Fig. 7.3 Original supporting program and layout of displacement monitor points

concrete is sprayed on the roadway surface with the thickness of 100 mm; and the row spacing of U29 arches is 800 mm.

7.2 Failure Mechanism of the Bearing Structure of Deep Soft Rock Roadways with High Stress

7.2.1 Monitoring of the Convergence and Deformation of Surrounding Rock

A corresponding monitoring program is designed for the deformation of the roadway surrounding rock on site. The monitoring points consist of Point O in the center, and Point A–F, M and N on the boundary of the roadway. The monitoring results are shown in Fig. 7.4.

The monitoring results in 198 days show the roadway surrounding rock has significant characteristics of large deformation, poor stability and long duration of deformation. Deformation occurs immediately after the completion of the roadway excavation. On the 10th day of monitoring, the convergence of the two sides reaches to 276 mm. In 20 days, the period of stable deformation of the surrounding rocks occurs. In 38 days, the surrounding rock deformation rate increases significantly.

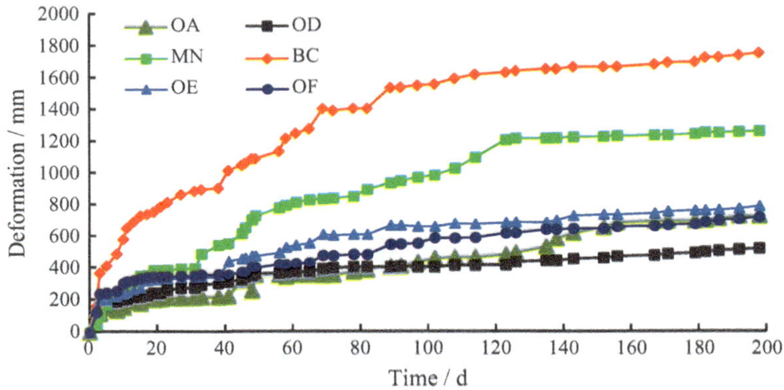

(a) Displacement-time curve of the roadway surface

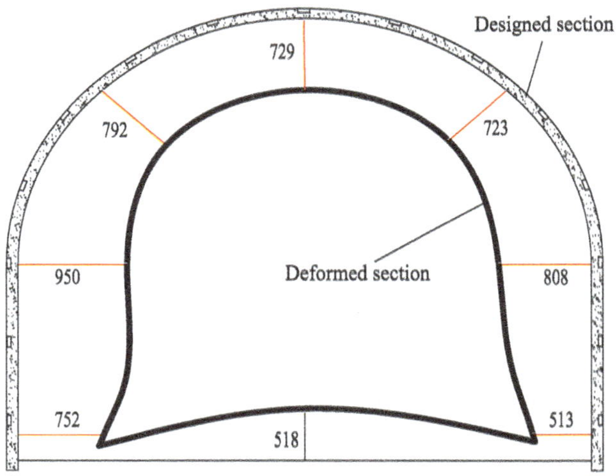

(b) Final shape of the roadway (unit: mm)

Fig. 7.4 Displacement curves and final shape of the roadway

After 145 days, the roadway deformation slows down, but continues. After the monitoring is completed, the entire displacement of the roadway surface shows the deformation characteristic of "the convergence of two sides > the roof settlement > the floor heave". The maximum deformation reaches to 1758 mm on two sides, the roof settlement is 729 mm, and the floor heave is 518 mm.

7.2.2 Detection on the Damage Range of the Surrounding Rocks

A borehole television is used to detect the damage range in order to analyze the internal destruction of the roadway surrounding rock. Three monitored sections are arranged in a typical roadway area with spacing of 20 m between each section and five detection drilling holes of Φ42 mm on each section. The detection boreholes are located on the vault, the left and right spandrel, and the left and right sides respectively. To facilitate the analysis of the results in each detection borehole, interpretation standards are established for the borehole detection in Zhaolou Coal Mine. With the standards, an agreement is made on the classification of the surrounding rock integrity, and the damage range of surrounding rock is determined in the roadway. As shown in Fig. 7.5, the black fillings in each borehole represent the observed location and extent of fragmentation on the inner wall of the borehole. After the drawing completion of the detection results in each hole, the damage zones of the surrounding rock is partitioned as severe, medium and minor, according to the extent of damage, from the inside to the outside.

The analysis shows:

(1) The surrounding rock damage is large in range and deep inside the surrounding rock. The average distance is 7.43 m (about 2.52 times of the roadway diameter) from the outside boundary of the minor damage zone to the chamber wall. It is 5.39 m (about 1.83 times of the roadway diameter) from the outside boundary of the medium damage zone to the chamber wall. It is 3.26 m (about 1.11 times

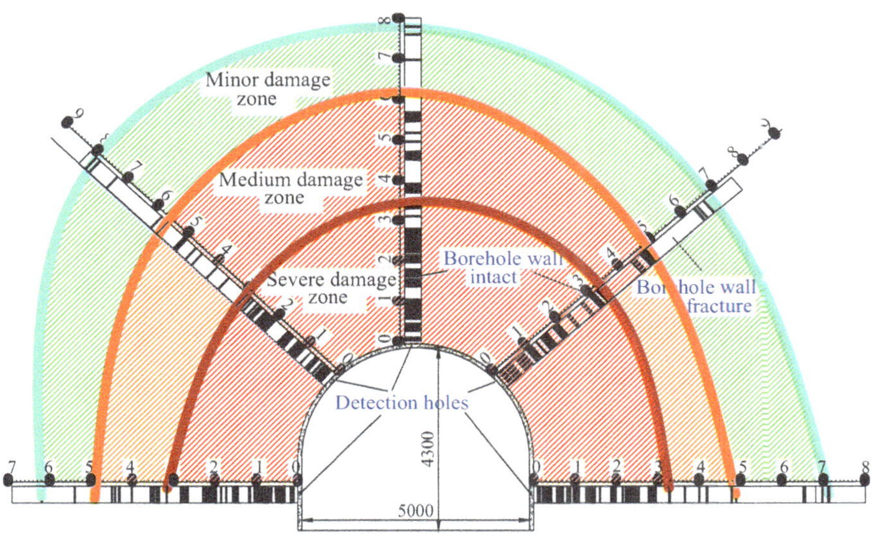

Fig. 7.5 Damage range of surrounding rock detection result

of the roadway diameter) from the outside boundary of the severe damage zone to the chamber wall.

(2) The damage zone is around the roadway and is nearly in an identical ring shape of the roadway.

(3) In the severe damage zone, the surrounding rock is fractured, and the cracks are extremely developed and are mainly opening cracks.

(4) In boreholes, the proportion of the clearly distinguished open cracks on the total number of cracks is in a trend of "on the roof > on the left side > on the right side" and "in the shallow surrounding rock > in the deep surrounding rock".

7.2.3 *Anchor Bolt Stress Monitoring*

In order to analyze the bolt stress state, three sections are set up for monitoring with five anchor bolt dynamometers installed on each section. The dynamometers are located on the left and right sides, the vault, and the left and right spandrel. The monitoring results are shown in Fig. 7.6.

The following conclusions can be drawn from the monitoring results:

The initial anchor bolt stress is small and the maximum of pre-tightening force is 50 kN. Within 15 days after the installation, the anchor uplifting force increases rapidly and then slows down. Around the 55th day, the force reaches its maximum, and then begins to decline. Up to the 198th day of observation, the force is 45 kN, 75 kN, 65 kN, 50 kN and 80 kN respectively on the left side and left spandrel, the vault, the right spandrel and right side; and it has declined by 66.2%, 44.1%, 59.4%, 66.7% and 50% comparing to its maximum value.

Combined with the results of drilling exploration, the analysis also shows in the post stage of the observation, the deformation of surrounding rock is too large; and the fragmentation of the surrounding rock is aggravated. That results in a further

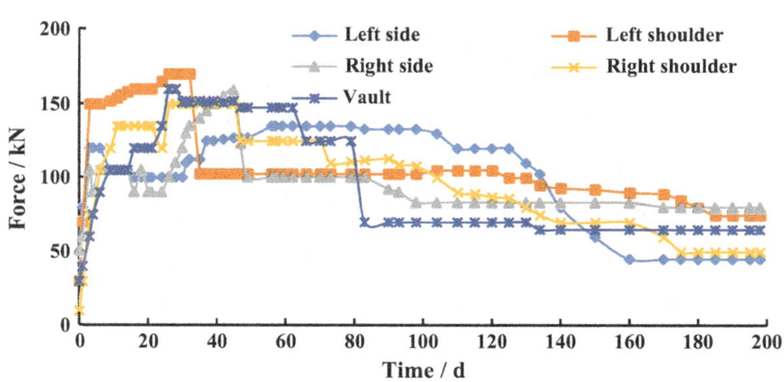

Fig. 7.6 Bolt force-time curve

increase of the damage range, anchor bolts being embraced generally in the severely damaged zone of surrounding rock; and therefore anchoring force is reduced or failed, and the anchor bolt fails to play its support potential.

7.2.4 Arch Stress Monitoring

A pressure gauge is used to monitor radial stress on the U29 arch. The monitoring sections are arranged with seven measuring points located on the left and right arch legs, the left and right skewbacks, the left and right spandrels, and the vault. The arrangement of the pressure gauge on site is shown in Fig. 7.7 and the arch monitoring results are shown in Fig. 7.8.

The analysis shows:

(1) The stress on some measuring points reaches to 60 kN within 7 days after the installation of the pressure gauge. The load continues to rise over time. The radial force rises most significantly on the vault, and reaches to 110 kN on the

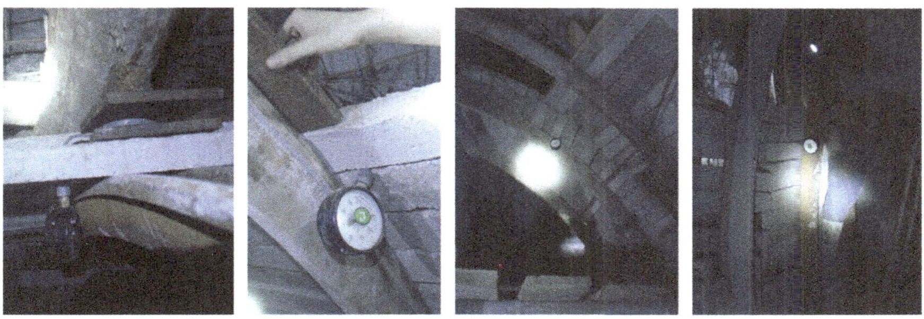

Fig. 7.7 Radial force monitoring on arch

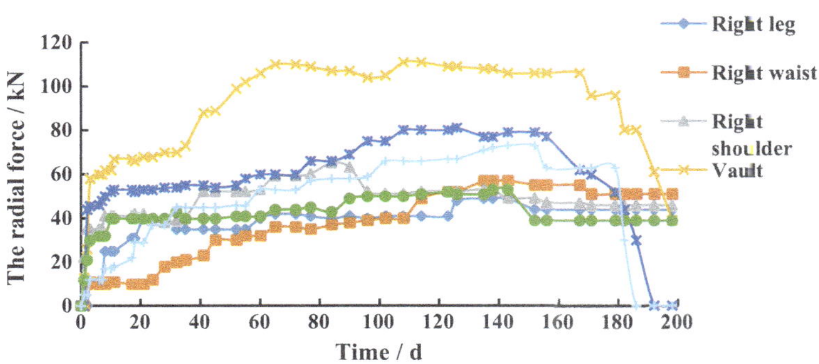

Fig. 7.8 Arch force-time curve

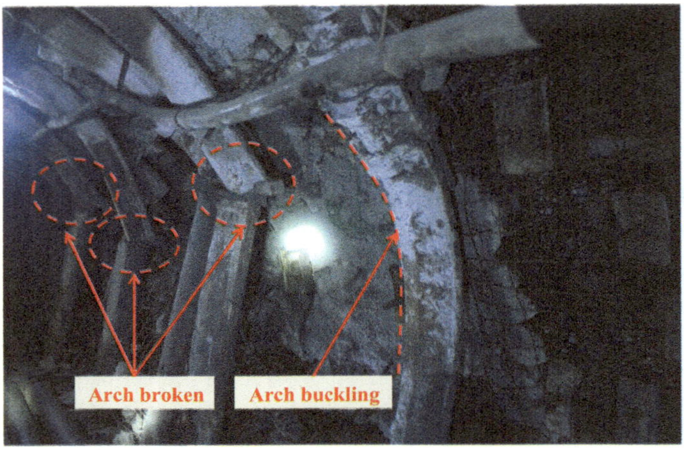

Fig. 7.9 Arch destruction

70th day. Between the 70th and the 150th days, arch radial stress is basically stable, with small increase or decrease only on individual measuring points. After the 150th day, the radial stress of the arch on some measuring points shows declining phenomenon; it declines to 0 kN at the measuring point on the left leg of the arch on the 186th day, and declines to 0 kN on the left spandrel on the 192th day; the clear buckling failure appears on the arch on site; and some arches are broken and lose their bearing capacity.

(2) The addition analysis on the monitoring data of the arch force shows the maximum radial load sum of the arch is 469 kN. The monitored arch has more uniform contact with the surrounding rock with the addition of pads and other special measures in the construction; and therefore, its bearing capacity is better exerted. In contrast, on the other arches without the measures, buckling and breaking phenomenon appear earlier as shown in Fig. 7.9. It is indicated that the bearing capacity of the U-steel arch is insufficient; its strength is low; its contact with surrounding rock is not uniform; and the potential of arch support is small, which cannot meet the stable control requirements of the surrounding rock of the deep soft roadway with high stress.

7.3 Control Mechanism of Surrounding Rock of Deep Roadways with High Stress

The authors have independently developed a test system for the high strength arch in underground engineering. With the system, full scale comparative laboratory experiments are conducted on the U29 and the SQCC arches of same size in the site of

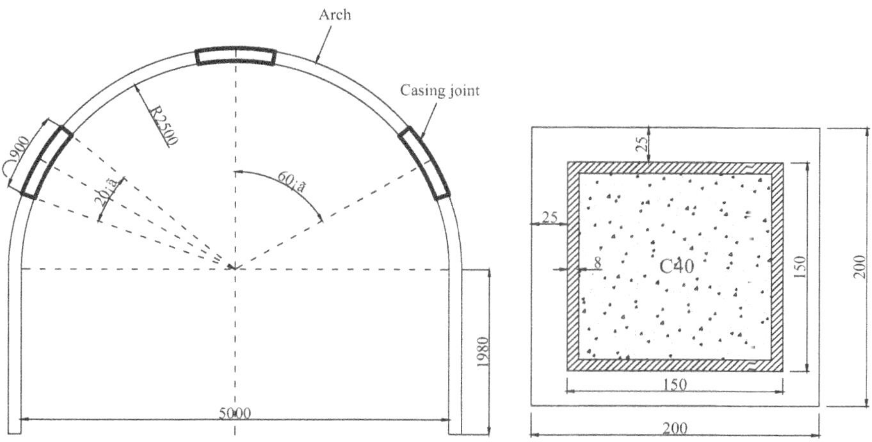

(a) Arch size (b) Section form of arch

Fig. 7.10 Size and section form of SQCC150 × 8 arch

Zhaolou coal mine. In the experiments, research is made on the stress, deformation and failure mechanism of the entire arch; and comparative analysis is made on the bearing capacity and the failure modes of different types of arches.

7.3.1 Laboratory Experiment Scheme

Square steel pipe of 150 × 8 mm (side length × wall thickness) is adopted for the SQCC arch (SQCC150 × 8 arch) with C40 concrete filled in the arch. The specific size parameters of the experimental arch are shown in Fig. 7.10. The U29 and the SQCC150 × 8 arches are in the same size. Each section of the U29 arch and the SQCC arch is connected by the clamps and casing joints respectively. The two types of arches have same diameters and similar cross-section steel content, and the uniform loading method is used.

7.3.2 Analysis on the Laboratory Experiment Results

7.3.2.1 Analysis on Failure Mode

SQCC150 × 8 arch has inconspicuous deformation and the thrust of each cylinder evenly increases over a longer stage after the experiment starts. As load continues to increase, the left and right sides of the arch are squeezed inward. When the total load of all cylinders reaches to 920 kN, the more obvious bending phenomenon occurs on the arch legs. When the load reaches to 1178 kN, the arch legs continue to be squeezed inward and the arch enters into its yield stage. When the load reaches to its maximum of 1287 kN, bending and deformation are obvious on the arch legs, particularly on its left leg. At the end of the experiment, the steel is not obviously deformation and the shape of the square steel cross-section substantially remains unchanged. All of that indicate the arch leg is still able to withstand a certain load, and the damage is not significant on the arch as a whole. The failure mode of SQCC150 × 8 arch after the test is shown in Fig. 7.11.

The entire U29 arch converges inward and deformed quickly after the test starts. As the load continues to increase and the total load of all the cylinders reaches to 205 kN, the arch legs converges inward and the settlement occurs on the vault. When the load reaches to its maximum of 598 kN, the arch legs are more obviously converges inward, particularly its left leg. As the load continues, the serious "Z" shape bent damage appears on the arch legs, and the cross-section is severely damaged. The total load has the sudden and sharp decline; and the entire arch becomes unstable and loses its bearing capacity. The failure mode of the U29 arch after the test is shown in Fig. 7.12.

Fig. 7.11 Failure mode of SQCC150 × 8 arch

Fig. 7.12 Failure mode of U29 arch

7.3.2.2 Bearing Capacity Analysis

Figure 7.13 shows the variation curves of the total load of the 9 groups of cylinders over time in the experiments on SQCC150 × 8 and U29 arches.

The analysis shows:

(1) The ultimate bearing capacity of the SQCC150 × 8 arch is 1287 kN, and that of U29 arch is 598 kN, and the former is 2.15 times of the latter. The SQCC arch has the higher bearing capacity than the traditional U-steel arch, providing the high-strength supporting reaction for surrounding rock.

(2) SQCC150 × 8 arch has the higher bearing capacity even after large deformation. After reaching the ultimate bearing capacity, with the continuous application of load, its bearing capacity decreases slowly. In contrast, the U29 arch has the low bearing capacity with the significant decrease stage of the curve. When its ultimate bearing capacity is reached, the flexural failure occurs on the arch legs suddenly. Its integrity is destroyed, and the bearing capacity declines drastically. In addition, SQCC arch has the better post bearing capacity than the U-steel arch, and it still has higher supporting reaction even with large deformation. However, the U-steel arch loses its stability after deformation occurs.

Fig. 7.13 Total load-time curve of arch

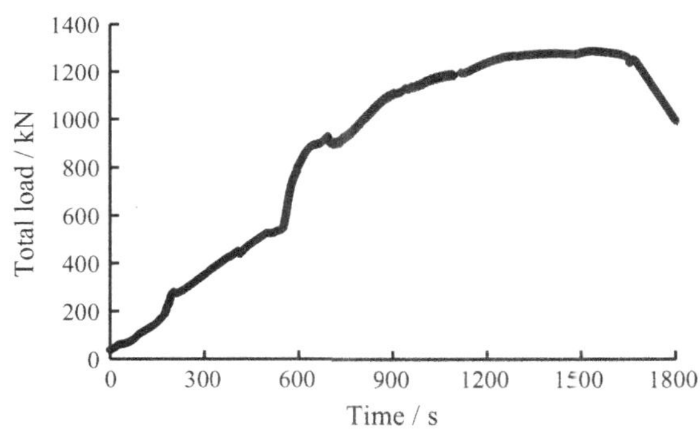

(a) Total load-time curve of SQCC150×8 arch

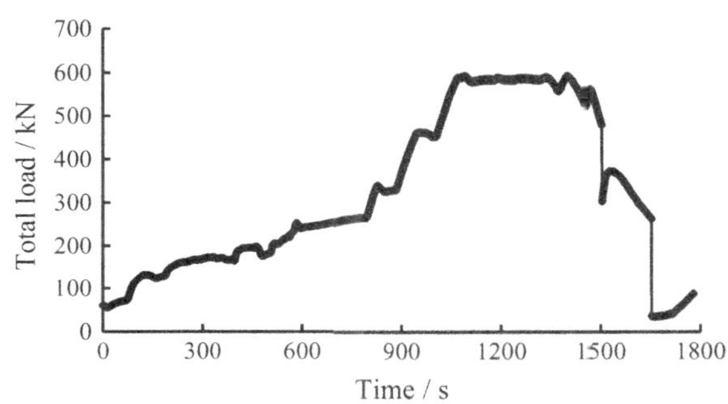

(b) Total load-time curve of U29 arch

7.3.3 *Numerical Comparative Experiment*

7.3.3.1 Experiment Scheme

The main factors influencing the stability of surrounding rock include supporting strength, geo-stress, and surrounding rock mechanical parameters. Numerical comparative experiment schemes are designed for three major categories in this section. The dimension of the roadway cross-section and the bolt-shotcrete support parameters as invariants. The field production requirements and the limitations of bolt models are also taken into account in the schemes. The researches are conducted on the mechanical properties of the supporting components and the surrounding rock control effect under the impact of the above factors.

The three major categories are the bolt-shotcrete support, the (i/4) U29 arch combined support and the (i/4) SQCC150 × 8 arch combined support; and their corresponding numbers are A, U_i and S_i, respectively, as shown in Table 7.1. In the table, $i = 1 - 8$, corresponds to 0.25–2 times of the arch strength, respectively; and the arch combined support schemes include the bolt-shotcrete support. The schemes are designed to study the influence factors of the support strength; they are based on geo-stress and surrounding rock mechanical parameters measured from the roadway of the second set of track downhill in Zhaolou Coal Mine as invariants and the arch strength as variable.

To study the influence factor on geo-stress, the measured surrounding rock mechanical parameters are considered as invariants and the geo-stress is considered as variable. The corresponding number is B_j, as shown in Table 7.2, where, $j = 1 - 7$, corresponding to 0.4–1.6 times of the measured geo-stress, respectively.

To study the influence factor on surrounding rock mechanical parameters, the elastic modulus E, the cohesion C and the internal friction angle φ are selected as the reference criteria, and the measured geo-stress is considered as the invariant, and the mechanical parameters of surrounding rock are taken as variables. The corresponding number is C_m, as shown in Table 7.3, where, $m = 1 - 7$, corresponding to 0.7–1.3 times of the surrounding rock mechanical parameters, respectively.

With the arch strength changed and the other factors remaining unchanged, the comparative analysis is made on the effect of the arch strength on the surrounding rock control. The specific scheme is shown in Table 7.1. 17 comparative schemes are

Table 7.1 Arch strength comparative scheme

No.	Scheme	Invariants	Remarks
A	Bolt-shotcrete support	The measured geo-stress	$i = 1 - 8$
U_i	The (i/4) U29 arch combined support	The measured surrounding rock mechanical parameters	
S_i	The (i/4) SQCC150 × 8 arch combined support		

Table 7.2 Geo-stress comparative scheme

No.	Degree of geo-stress	Geo-stress/MPa	Invariants
B1	0.4	10	Arch strength U4, S4 The measured surrounding rock mechanical parameters
B2	0.6	15	
B3	0.8	20	
B4	1.0	25	
B5	1.2	30	
B6	1.4	35	
B7	1.6	40	

Table 7.3 Surrounding rock mechanical parameters comparative scheme

No.	Degree of surrounding rock parameters	Elastic modulus *E*/GPa	Cohesion *C*/MPa	Internal friction angle $\varphi/°$	Invariants
C1	0.7	1.12	0.42	19.6	Arch strength U4, S4 The measured geo-stress
C2	0.8	1.28	0.48	22.4	
C3	0.9	1.44	0.54	25.2	
C4	1.0	1.6	0.6	28	
C5	1.1	1.76	0.66	30.8	
C6	1.2	1.92	0.72	33.6	
C7	1.3	2.08	0.78	36.4	

designed in the 3 categories. Where, scheme U4 represents the U29 arch combined support and scheme S4 represents the SQCC150 × 8 arch combined support.

(1) Geo-stress Comparative Scheme
With the geo-stress changed and the other factors remaining unchanged, the comparative analysis is made on the effect of the geo-stress on the surrounding rock control. The specific scheme is shown in Table 7.2. 14 comparative schemes are designed in the 2 categories. Where, scheme B4 represents the measured geo-stress.

(2) Surrounding Rock Mechanical Parameters Comparative Scheme
With the surrounding rock mechanical parameters changed and the other factors remaining unchanged, the comparative analysis is made on the effect of the surrounding rock mechanical parameters on the surrounding rock control. The specific scheme is shown in Table 7.3. 14 comparative schemes are designed in the 2 categories. Where, scheme C4 represents the measured surrounding rock mechanical parameters.

7.3.3.2 Model and Evaluation Index Establishment

(1) Model Establishment

In order to obtain the law with more universal significance, the model is simplified according to the actual situation on site. The rock layer is simplified to the horizontal, and no surrounding rock is distinguished by lithology. The model dimension is 40 m × 40 m × 1.6 m (height × width × thickness). The constraint is carried out from x, y and z three directions at the bottom and from x and y directions at the two sides, in front and in rear (as shown in Fig. 7.14).

(2) Parameter Selection

Mohr–Coulomb criterion is used on the surrounding rock. The specific physical and mechanical parameters are shown in Table 7.4. The arch dimension and mechanical parameters are shown in Table 7.5.

The parameters of U-steel and SQCC arches are determined according to their corresponding steel mechanical properties, as shown in Table 7.5. The concrete is C40 in the core of the square steel.

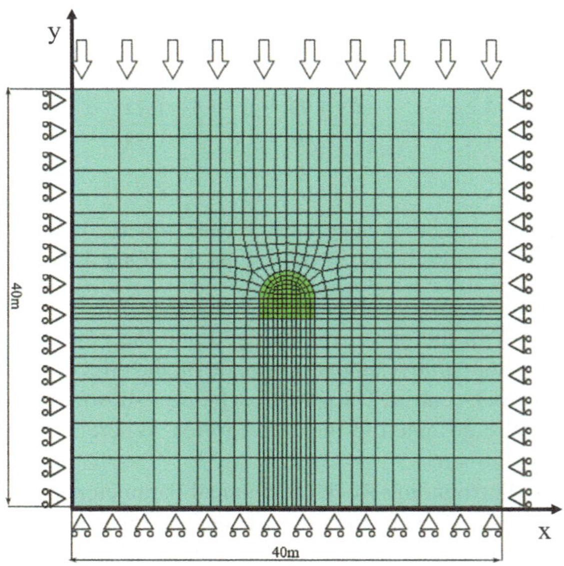

Fig. 7.14 Numerical calculation model

Table 7.4 Mechanical parameters of anchor

Support component	Component dimension/mm	Poisson ratio	Yield strength σ_s/MPa	Ultimate strength σ_b/MPa	Elastic modulus E/GPa
Bolt	Φ22 × 2400	0.3	500	700	195

Table 7.5 Mechanical parameters of arch

Arch type	Elastic modulus *E*/GPa	Poisson ratio μ	Yield strength σ_s/MPa	Ultimate strength σ_b/MPa
U29	197	0.3	335	533
SQCC150 × 8	206	0.3	235	510

(3) Evaluation Index Establishment

Two quantitative evaluation indices are established, which are vault displacement control rate and bolt strength utilization rate. The surrounding rock control effect and the supporting component mechanical properties under the influence of the different factors are comparatively analyzed.

The vault displacement control rate $\delta_{\rm Dn}$

$$\delta_{\rm Dn} = \frac{D_0 - D_n}{D_0} \times 100\% \tag{1}$$

where, D_0 is the vault displacement in the same experiment scheme with the bolt-shotcrete support, mm; and $D_{\rm n}$ is the vault displacement in the same experiment scheme with the combined support, mm.

The anchor bolt strength utilization rate η

$$\eta = \frac{\sigma_n}{\sigma_{\rm u}} \times 100\% \tag{2}$$

where, $\sigma_{\rm n}$ is the maximum stress on anchor bolt in same experiment scheme with combined support, Mpa; and $\sigma_{\rm u}$ is the bolt ultimate stress, MPa.

7.3.4 Results Analysis

7.3.4.1 Results Analysis on the Arch Strength Comparative Schemes

The calculation results of the arch strength comparative schemes are plotted in Figs. 7.15 and 7.16.

The comparative analysis shows:

(1) With scheme U4, which refers to the U29 arch combined support, the minimum vault displacement is 406.73 mm and the displacement control rate is 20.3%. With scheme S4, which refers to the SQCC150 × 8 arch combined support, the minimum vault displacement is 143.8 mm and the displacement control rate is 72.0%. The displacement control rate of the latter is 3.46 times that of the former. The SQCC150 × 8 arch combined support has the remarkable surrounding rock control effect.

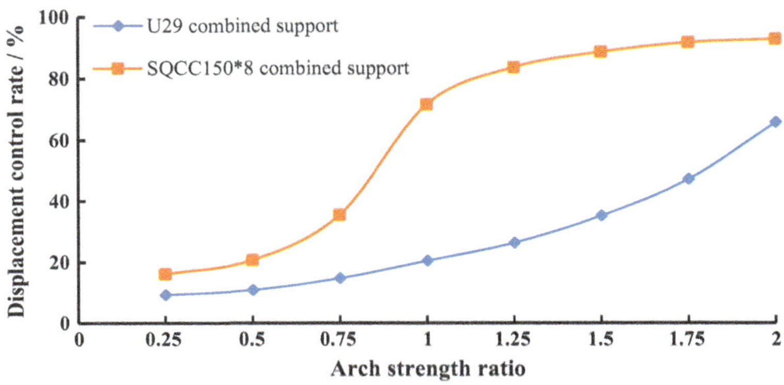

Fig. 7.15 Vault displacement control rate with different arch strength

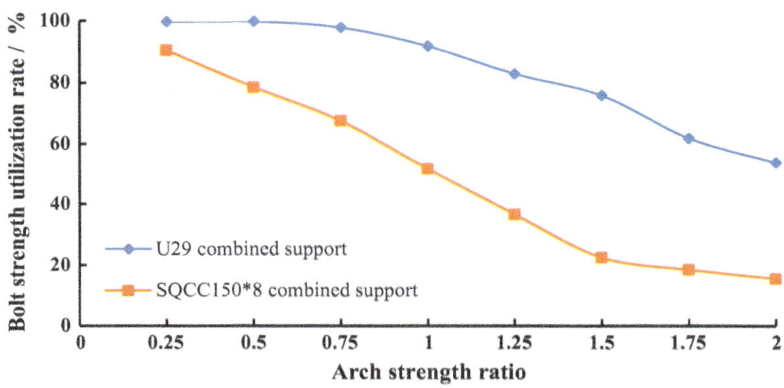

Fig. 7.16 Anchor bolt strength utilization rate with different arch strength

(2) Along with the increase of arch strength, displacement control rate increases and surrounding rock control effect gradually becomes better. With the U29 arch combined support, the vault displacement control rate shows the similar exponential increase with the arch strength; and that indicates the improvement of the U-steel arch strength has a good effect on surrounding rock control under this geological conditions. With the SQCC150 × 8 arch combined support, the vault displacement control rate substantially shows the exponential increase as the arch strength increases by 0.25–1 times; and as the arch strength further increases, the growth of displacement control rate slows down. That indicates the SQCC150 × 8 arch combined support has the remarkable control effect on surrounding rock under this geological conditions; and the continuous increase of arch strength could reduce its utilization rate.

(3) Along with the increase of arch strength, bolt strength utilization rate decreases and the rock stress also decreases gradually on the bolt. In the U29 arch combined

support scheme with arch strength ratios of 0.25 and 0.5, the bolt utilization rate reaches to 100% which means the bolt is broken. In the SQCC150 × 8 arch combined support scheme, the bolt strength utilization rate is significantly low, its average is only 48.3% of that in the U29 arch combined support. The SQCC150 × 8 arch combined support has the good supporting strength reserve and could achieve better coupling with the supporting components.

(4) The combined support of bolt-shotcrete + arch is a good solution to the issue of surrounding rock control in deep roadways with high stress. As the arch strength increases, its effect on the surrounding rock displacement control is improved. With the high strength and the high post bearing capacity, the SQCC arch produces the better effect than the U-steel arch does on surrounding rock control.

7.3.4.2 Result Analysis on Geo-Stress Comparative Schemes

Calculation results of geo-stress comparative schemes are plotted in Figs. 7.17, 7.18 and 7.19.

(1) Along with the increase of geo-stress, the vault displacement increases and the displacement control rate decreases gradually under the conditions of the two combined supports. The effect of surrounding rock control is better under the SQCC150 × 8 arch combined support than the U29 arch combined support. The average displacement control rate of the former is 5.1 times higher than that of the latter. Particularly in scheme B7 which has the maximum geo-stress, the vault displacement control rate is only 17.3% under the U29 arch combined support and it still could reach 63% under the SQCC150 × 8 arch combined support.

(2) Along with the increase of geo-stress, the bolt strength utilization rate increases gradually. The bolt strength utilization rate increases from 37 to 69% under scheme S4 (the SQCC150 × 8 arch combined support); however, the bolt strength utilization rate increases from 69 to 100% under scheme U4 (the U29 arch combined support). In schemes B5–B7, all the bolt strength utilization rates reach 100%, which means the bolt is broken. The average bolt strength utilization rate is 1.74 times higher under the U29 arch combined support than under the SQCC150 × 8 arch combined support. That indicates the bearing capacity of the U29 arch is much lower than that of the SQCC150 × 8 arch; and as the result, the corresponding bolts will bear more surrounding rock pressure and be failed easily. The improvement of arch strength can reduce the bolt utilization rate to some extent and increase support safety reserve. SQCC arches could play a good role in this respect.

(3) Along with the increase of geo-stress, the maximum arch stress increases gradually, and the maximum arch stress increases linearly under the SQCC150 × 8 arch combined support. The maximum arch stress increases faster under the U29 arch combined support in schemes B1–B3. The stress reaches its limit in

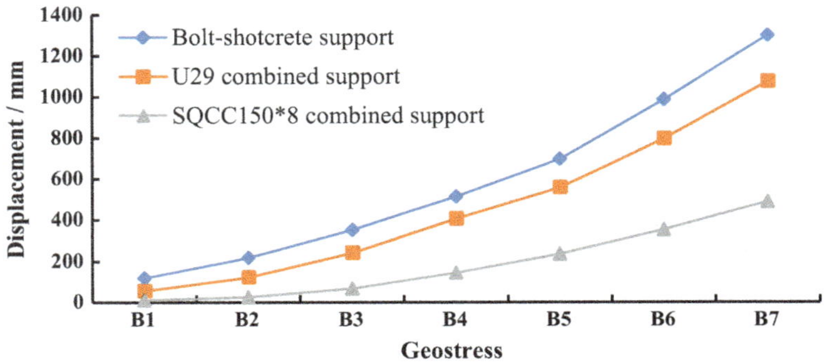

(a) Variation curve of vault displacement

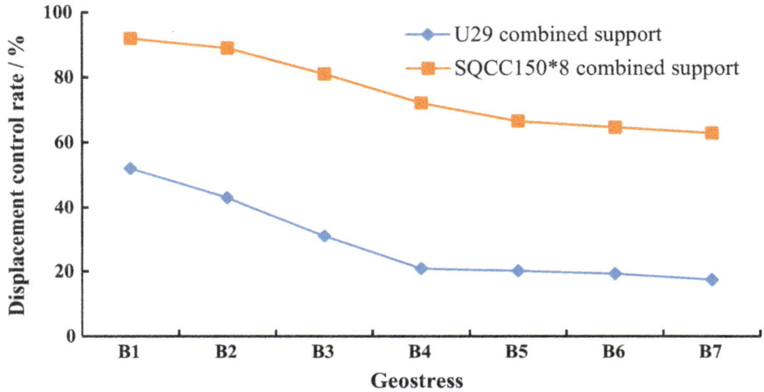

(b) Variation curve of vault displacement control rate

Fig. 7.17 Variation of vault displacement and control rate

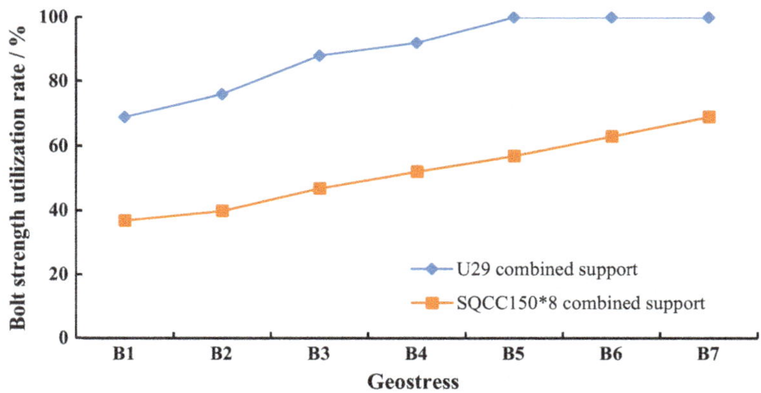

Fig. 7.18 Variation of bolt strength utilization rate

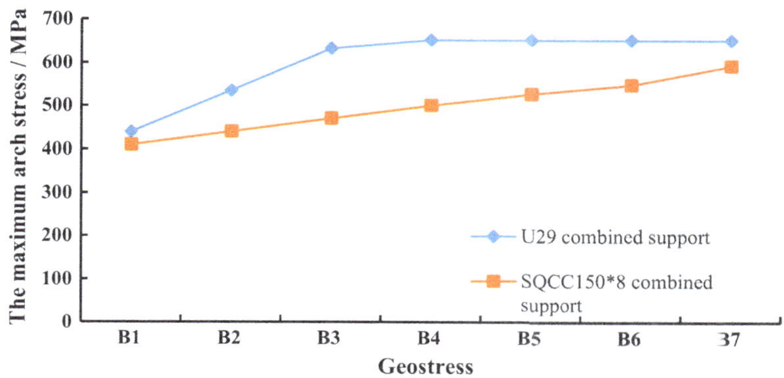

Fig. 7.19 Variation of bolt the maximum arch stress

schemes B3–B7, meaning serious strength failures have occurred. In contrast, the stress on the SQCC150 × 8 arch has not yet reached the limit even in the scheme B7 which has the maximum geo-stress, indicating SQCC150 × 8 arch has the good bearing capacity.

7.3.4.3 Result Analysis on Comparative Schemes of Surrounding Rock Mechanical Parameters

The calculation results of the comparative schemes of surrounding rock mechanical parameters are plotted in Figs. 7.20, 7.21 and 7.22.

The analysis shows:

(1) Along with the increase of surrounding rock mechanical parameters, both displacement and displacement control rate on the vault show the decreasing trend in all support schemes. That is mainly because the bolt-shotcrete support produce the much better control effect and reduce the role that the arch plays on surrounding rock control under the improved surrounding rock conditions. The influence of the increase of mechanical parameters of surrounding rock in schemes C5–C7 on the vault displacement control rate under different support conditions becomes smaller. The average vault displacement control rate in the SQCC150 × 8 combined support is 2.5 times of that in the U29 arch combined support. That indicates the former has more obvious advantages than the later.
(2) Along with the increase of surrounding rock mechanical parameters, the bolt strength utilization rate decreases gradually. The utilization rate decreases from 72.1 to 22% in the SQCC150 × 8 arch combined support and from 100 to 65% in the U29 arch combined support. The latter is 1.4–3.0 times that of the former, indicating the bolt strength utilization rate is higher and the safety reserve is smaller in U29 arch combined support on the surrounding rock with lower strength, especially on weak and broken rock mass.

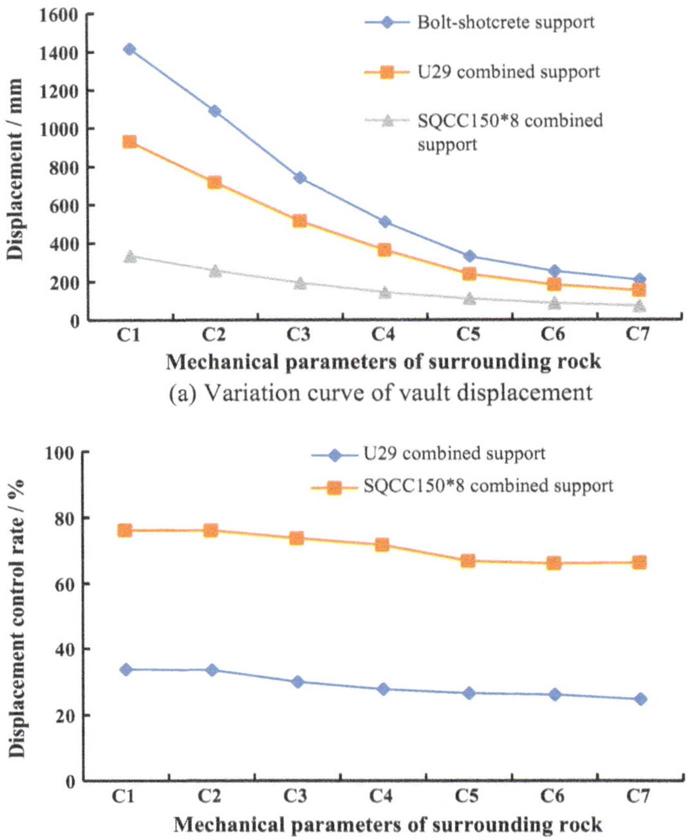

(a) Variation curve of vault displacement

(b) Variation curve of vault displacement control rate

Fig. 7.20 Variation of displacement control rate

(3) Along with the increase of surrounding rock mechanical parameters in schemes C1–C4, the maximum arch stress declines gradually. As its ultimate stress is reached, U29 arch is severely damaged and even broken. As the surrounding rock conditions improves gradually, the arch stress declines gradually. The average maximum stress on the U29 arch is 1.2 times that of the SQCC150 × 8 arch. The stresses are the closest on the two arches in scheme C7, indicating better surrounding rock conditions can make the self-bearing structure easily to be formed. The effect of the passive support form is reduced. The stress is not high on both types of the arches. When the surrounding rock mechanical parameters decrease, the commonly used U-steel arch support can no longer meet the need of the surrounding rock control.

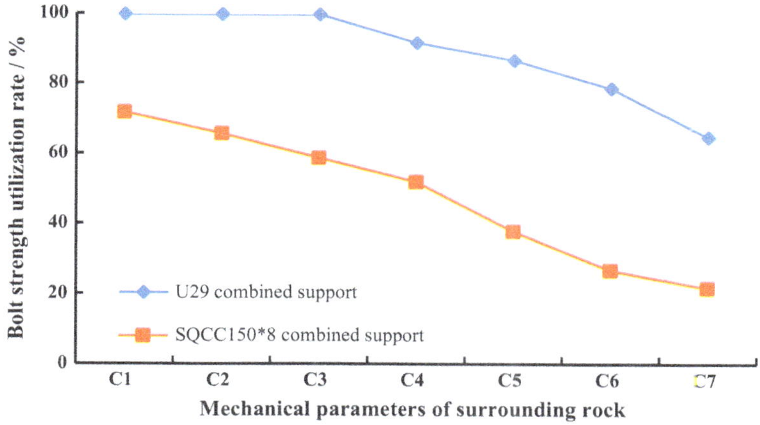

Fig. 7.21 Variation of the use rate of bolt strength

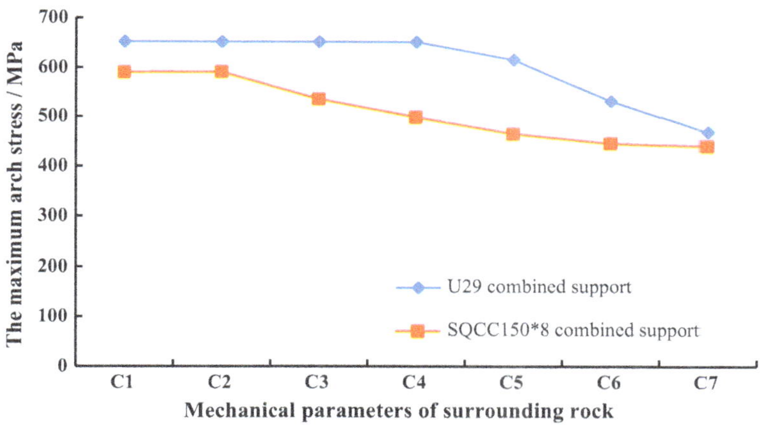

Fig. 7.22 Variation of the maximum arch stress

7.4 Engineering Practice of Deep Roadways with High Stress

7.4.1 Scheme Design

The second set of track downhill mentioned in Sect. 7.1 is selected as the roadway test to conduct the comparative tests on surrounding rock control effect with both the U29 and the SQCC arches, respectively. The designed cross-section of SQCC arch is shown in Fig. 7.23 with the arch row distance of 1 m. The U29 arch support parameters are consistent with that of SQCC arch in test section; and the parameters

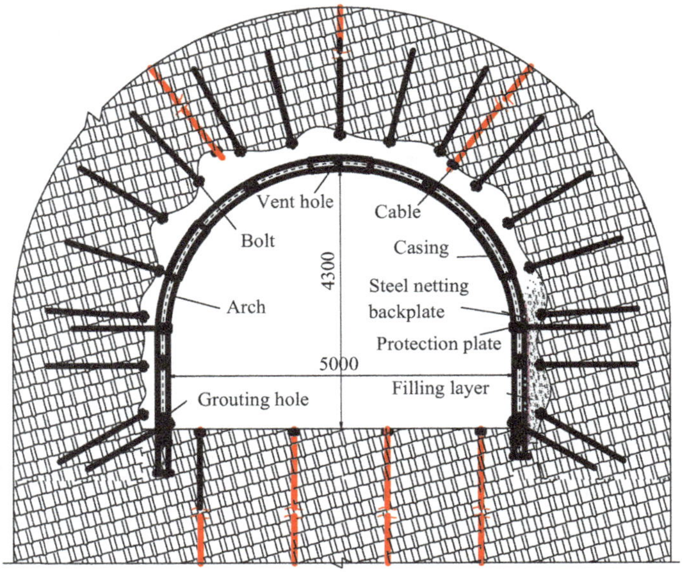

Fig. 7.23 Sectional drawing of SQCC support system

of the bolt-shotcrete support are the same as that in the original scheme. Meanwhile, anchor cable is installed on the floor in both schemes in this test to control the floor heave. The cable size is Φ22 × 6200 mm with inter-row spacing of 1500 × 1600 mm.

Five monitoring cross-sections are arranged in each of the two schemes in this test. Pressure gauges are used to monitor arch radial stress, mainly on the vault, the spandrel, the springing and the skewbacks. Convergence rulers are used to measure the convergence of the roadway and the arch, mainly for monitoring the vault settlement, the convergence on the two sides and the floor heave. The monitoring equipment distribution on each cross-section is shown in Fig. 7.24.

Installation of the monitoring equipment and construction process of SQCC support system are shown in Fig. 7.25.

7.4.2 Analysis of Monitoring Results

Figure 7.26 shows the radial force and the deformation of the typical arch positions over time on the second monitoring cross-section. "SQCC-D-6#" and "SQCC-F-7#" are taken as examples. "D" represents the displacement, "F" represents the radial force, "6#" represents 6# measuring point in Fig. 7.26, and so on. Figure 7.27 shows the comparison of the final stress and the deformation of the SQCC arch and the U29 arch after the excavation of 157 days.

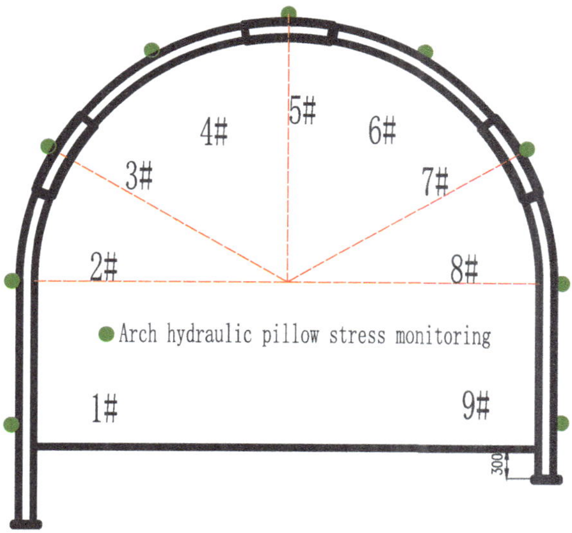

Fig. 7.24 Monitoring sectional drawing of SQCC support system

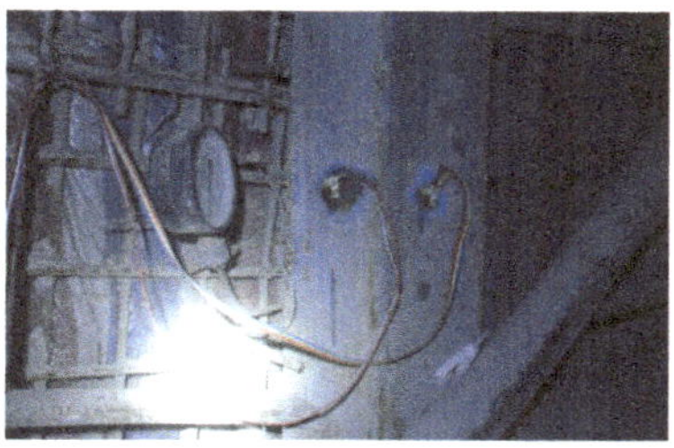

(a) Installation of radial pressure gauges

(b) Construction of arches

Fig. 7.25 Field application of SQCC

The analysis shows:

(1) As shown in Fig. 7.27, the obvious roadway deformation occurs on the two spandrel on the test section of the SQCC arch with the maximum of 29.7 mm and the inconspicuous vault settlement. In contrast, the maximum deformation of 167.3 mm occurs on the upper part of the left straight leg of on the test section of the U29 arch with the obvious vault settlement, and the deformation on the two spandrel is large as well. On the 157th day of monitoring, the average deformation of the roadway surrounding rock is 15.3 mm on all the monitoring points on the test section of the SQCC arch; five points of them show the deformation even less than 15 mm; the overall arch deformation is small; and its deformation is only 21.2% of that on the test section of the U29 arch. The

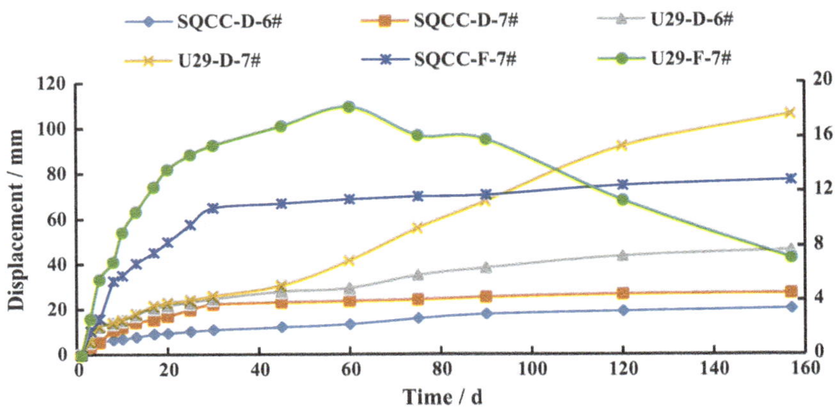

Fig. 7.26 Curves of roadway surface and radial pressure of some monitoring points

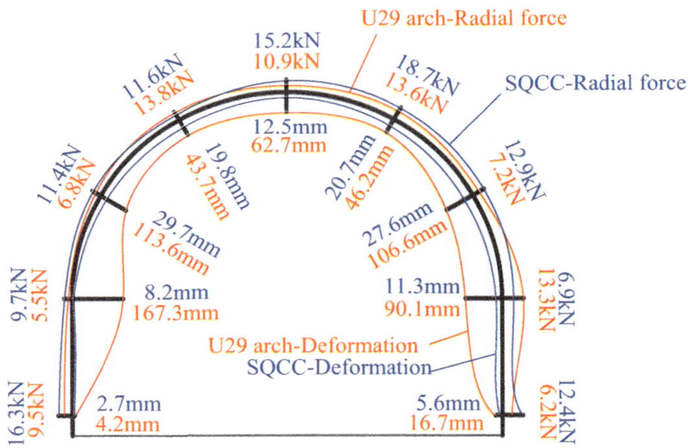

Fig. 7.27 Monitoring results of the deformation and the radial force of the arch

monitoring results at monitoring point 7# on the U29 arch in Fig. 7.26 show the continuation of deformation still on the test section roadway of the U29 arch.

(2) On the 157th day of monitoring, the average radial stress at each measuring point is 12.8 kN and 9.6 kN respectively on the SQCC and the U29 arches; and the difference is small. The field observation indicates that some arches are partially yielded and broken on the test section of the U29 arch. The U29 arch at 7# measuring point in Fig. 7.26 is taken as an example; the deformation is 42.2 mm on the 60th day of monitoring and then the deformation accelerates; the arch radial stress starts to decline from 18.3 to 7.2 kN at the end of monitoring in this location; and the deformation increases to 106.6 mm during this period.

Fig. 7.28 Surrounding rock control effect of the SQCC support system

In contrast, the roadway surrounding rock is well controlled and stabilized on the test section of the SQCC arch. No similar phenomenon occurred.

Surrounding rock control effect of the SQCC support system on site is shown in Fig. 7.28.

7.5 Chapter Summary

Taking Zhaolou Coal Mine as the engineering background, the full scale laboratory experiments on the SQCC and the traditional U29 arches are carried out in this chapter. The experiment shows that the ultimate bearing capacity of the SQCC arch is 2.15 times that of the U29 arch with similar steel content. In this chapter, the quantitative evaluation index is established; and the control mechanism of surrounding rock and the stress characteristics of supporting members are compared and analyzed under the different influence factors. The numerical test results show the SQCC arch combined support can provide the large supporting force to ensure the overall effectiveness of the support system; the SQCC150 × 8 arch combined support is better than that the U29 arch combined support in surrounding rock control. The field test shows that the stability of roadway surrounding rock is effectively controlled in the test section of the SQCC arch, and the good control effect of the SQCC arch support is verified on surrounding rock.

The manufacturer's authorised representative in the EU is Springer Nature Customer Service Centre GmbH, Europaplatz 3, 69115 Heidelberg, Germany. If you have any concerns regarding our products, please contact ProductSafety@springernature.com

Printed and bound by CPI Group (UK) Ltd, Croydon, CR0 4YY

30/07/2026

02181699-0001